$MART POWER

An urban guide to renewable energy and efficiency

Smart Power

$MART POWER

An urban guide to renewable energy and efficiency

William H. Kemp

AZTEXT PRESS

Tamworth, Ontario
www.aztext.com
cam@aztext.com

Distributed By:
Hushion House Publishing Limited
36 Northline Road, Toronto, Ontario Canada M4B 3E2
Telephone: (416)-285-6100 E-mail: jbeau@hushion.com

$mart Power

Library and Archives Canada Cataloguing in Publication

Kemp, William H., 1960-
     Smart power : an urban guide to renewable energy and efficiency / William H. Kemp.

Includes index.

Letter "s" in the word "smart" at beginning of title represented by a dollar sign.

ISBN 0-9733233-1-0

     1. Dwellings--Energy conservation. I. Title.

TJ163.5.D86K45 2004   644   C2004-905456-2

Disclaimer:
"The installation and operation of renewable energy systems involves a degree of risk. Ensure that all proper installation, regulatory and common sense safety rules are followed. If you are unsure of what you are doing, STOP! Seek skilled and competent help by discussing these activities with your dealer or electrician.

Electrical systems are subject to the rules of the National Electrical Code™ in the United States, and the Canadian Electrical Code™ in Canada. Wood and pellet stoves are also subject to national and local building code requirements.

In addition, local utility, insurance, zoning and many other issues must be dealt with prior to beginning any installation. A word to the wise, "People don't plan to fail, they just fail to plan." Nowhere is this more apparent than when installing renewable energy systems.

The author and publisher assume no liability for personal injury, property damage, consequential damage or loss, including errors and omissions, from using the information in this book, however caused.

The views expressed in this book are those of the author, and do not necessarily reflect the views of contributors or others who have provided information or material used herein.

Printed and bound in Canada on recycled stock using vegetable based inks.

Table of Contents

Table of Contents

Preface

The majority of the population don't give much thought to energy. Sure, everyone gripes a bit (maybe a lot) when the cost of electricity or gasoline rises for the tenth time this year. But do most people really ever think about energy? Where does it come from and what impact does it have on the family budget, the environment, or our children's future?

Most of us are so busy trying to stay afloat in our hectic lives that these questions only surface as polite dinner discussions. Nevertheless, every once in a while someone stops to ponder these issues. Perhaps that's why you've picked up *$mart Power, An Urban Guide to Renewable Energy and Efficiency.*

Within the covers of this handbook, you will be provided with the background information, purchasing data, and step-by-step instructions to allow you to:

- stretch your heating and utility dollar;
- make the electrical meter run backwards by generating your own electrical power;
- provide energy for domestic hot water, space heating, and electricity from independent and clean sources;
- worry just a little bit less about acid rain, ozone depletion, airborne smog, nuclear waste, or world oil supply problems.

Whether you are just curious or an industry expert, if these issues are on your mind then *$mart Power* is for you.

We will discuss how to stretch your energy dollar, doing much more with less. Step-by-step guidance and easy-to-understand instruction will help pave the way to better energy management and renewable energy production, whether you own an estate in the suburbs or rent a small apartment downtown.

We will guide you to a new relationship with the environment that lets you consume energy without worrying about the effects of dumping today's pollution on tomorrow's children. Unlike fossil fuels or nuclear energy, renewable sources are just that: renewable. No matter how much energy we capture and use, there will always be more for everyone else.

And most important of all, *$mart Power* will show you how to do it economically and do it now!

Acknowledgements

A book of the magnitude of *$mart Power* requires the technical knowledge and access to information from many individuals and companies, not to mention the patience of family while writing and editing the text.

It is customary to identify those contributors who have helped with technical data required to put this book together and make sure the information is the most up to date.

I am indebted to, and would like to thank, Sean Twomey and the delightful staff at Arbour Environmental Shoppe; Michael Reed of Array Technology Inc. for information on tracking solar mounts; Mike Bergey of Bergey Wind Power for technical publications and photographs on wind power systems; Don Bishop of Benjamin Heating Products for information regarding dual energy heating boilers; Christine Paquette of the Biodiesel Association of Canada for assistance and review on the Biofuels Chapter; Liujin, Les, Jose and Steve from Carearth Inc. for installing a great solar thermal system on my house and providing information and photographs; Dr. David Suzuki, Chair, Jose Etcheverry, David Suzuki Foundation for endorsing this book and for mentoring me; Vanessa Percival at Embers; Nicole and Aidan Foss for being excellent critics and mentors of world energy policy; Paul Gipe   for his assistance with urban wind turbines; George Peroni at HydroCap Corporation for information regarding hydrogen gas reclaiming devices for batteries; General Motors Corporation for their photographs of hybrid and hydrogen vehicles; Ross and Kathryn Elliot of Homestead Building Solutions for help with the home air leakage tests; Patrice Feldman at Morning Star Corporation for her quick response regarding PV controller design and installation; Rajindar Rangi and Tony Tung of Natural Resources Canada, Energy Technology Branch for guidance and providing access to reams of technical bulletins and contacts; Gerald Van Decker of Renewability Energy Inc. for assistance with the waste water heat recovery system; John Supp of South West Wind Power for the great site photographs and micro-wind technical data; Govindh Jayaraman of Topia Energy Inc.; Toyota   Motor Company for their photographs and technical data regarding hybrid vehicles; Ian Micklethwaite of Warren Publishing Co. Ltd.; Pam Carlson of Xantrex Technology Inc. for the tremendous support of technical documentation, photographs and the nice Xantrex coffee mug!; and lastly, Stefani Kuykendall of Zomeworks Corp. for information on fixed PV mounting hardware. (If I have missed anyone, please accept my appologies and understand that your assistance was truly appreciated!)

A huge thank you must also be given to Joan McKibbin my editor. Joan worked tirelessly under very demanding deadlines and also stopped me from writing too many run-on sentences. I am surprised that she only needed a cup of tea and a hot bath after pulling our last "all nighter". A lesser person would have required something a bit more potent!

I am also indebted to those who have had to work directly with me in the preparation of the text. A big thanks goes out to: Carol McGregor and Cam Mather for their excellent line drawings and graphics, not to mention having to work under Bill's demanding time lines; Christina Dunn, Lorraine Kemp, Katie Mather for their great help behind the camera and proving the old adage that a picture is worth a thousand of my words. (All images without photo credits by them); To Karen and Jamie Wilson, Hillary Houston and Raymond Lebfrevre, The Adams, Lorraine Kemp and Michelle and Cam Mather for letting me invade their privacy to extract the photographs of R.E. systems throughout the book.

And of course thanks to Lorraine Kemp and Michelle Mather (and Katie and Nicole) for putting up with Cam Mather and me during our single-minded focus on writing and producing this book, using **100% Renewable Energy!** (Aztext Electronic Publishing Ltd. is available for your needs; contact cam@aztext.com.)

Cover Collage Acknowledgements

We'd like to thank the following individuals and corporations for allowing us to use their images on the cover of "$mart Power".

The wind turbines are courtesy of Vision Quest Wind Electric www.visionquestwind.com

The rider and trailer are courtesy of Burley Design Cooperative, www.burley.com

The sporty red "Smart Car" is courtesy of Smart™ gmbh www.mercedes-benz.ca

The photo of "Southfield House", owned by Peter Richards, was taken by Bob Swartman of Solcan Ltd., www.solcan.com

About the Author

William Kemp

As a consulting electronics/software designer, Bill develops high performance embedded control systems for low environmental impact hydroelectric utilities worldwide. He is actively involved in the development and advocacy of green-technologies including; renewable energy heating, energy efficiency, sustainable development as well as photovoltaic, micro-hydro and wind electric systems. Bill advises on grid-intertie and off-grid domestic systems and is a leading expert in small and mid-scale (<20MW) renewable energy technologies. He is the author of the best selling book The Renewable Energy Handbook for Homeowners and has just completed his second book titled $mart Power; an urban guide to renewable energy and efficiency. In addition he has published numerous articles on small-scale private power and is the chairman of an electrical safety standards committee with the Canadian Standards Association. He and his wife Lorraine, live off the electrical grid on their hobby/horse farm in eastern Ontario. Visit Bill's website at www.balancetoday.ca.

Environmental Stewartship

In our private lives, the principals in Aztext Press try to minimize our impact on the planet, living off the grid and producing our electricity through renewable energy sources and looking at all facets of our lives, including our personal transportation, heating, cooling, food choices, etc.

We continue to evaluate our books to try minimize their impact, and pick the right mix of recycled stock that still allow photographs to be clear and crisp and offer a quality product to motivate our readers.

In the last 10 years we have planted over 2,000 trees, and each year make a commitment to plant more and nurture existing trees so they thrive and maximize their potential to act like the lungs of the planet and clean the air.

$mart Power
Dedication

Six solar vehicles with their crews of students from Canadian universities were to drive 1,060 miles (1,700 km) from Windsor on the U.S.-Canada border to Québec City, visiting numerous towns and villages on the way. This unique event was timed to coincide with the anniversary of the disastrous 2003 power blackout, which left much of eastern North America in darkness. The inter-university cooperation in developing and racing these advanced technology vehicles strengthens the calibre of their research as well as showcasing clean energy technologies to the public (Figure 6-10). Many of the advanced designs used in these cars make their way through laboratories until they surface as highly energy-efficient products used in day-to-day life.

The overcast sky could not stop the energy from flowing in either the solar racers or their sixty student drivers and support technicians. With the cars pulling out in a high-tech convoy consisting of a lead vehicle, the solar racers, and a chase truck, you could feel the excitement and enthusiasm as the teams talked back and forth on crackling radios.

University of Toronto racer "Faust II" was en route, driving like the wind, when it was involved in a fatal accident, killing 21 year old Andrew Frow.

The radios were silenced after that fateful day.

As with all great human endeavours, there is a cost that pioneers often pay in making the way easier for the rest of society. Andrew paid that price and will be remembered as someone committed to advancing technology as he worked to move humanity towards cleaner and sustainable transportation.

With time to heal, the enthusiasm of youth will prevail and Andrew's efforts will not have been in vain.

Godspeed, Andrew.

1
Introduction

Energy is the life breath of modern society. Without energy, almost every event we take for granted comes to a screeching halt. If you happened to live on the eastern side of North America during the blackout of 2003, you will acutely understand how life without electricity changes everything. For a few dozen hours, modern life stopped— *completely*: people stuck in elevators; no building or traffic lights; traffic gridlocks; no air conditioning, cell phones, computers, television or email. The entire world appeared to go black.

North American dependence on electrical energy and fossil fuels is, per capita, the highest in the world. In the early 1900's the United States was the world's largest producer of oil. Over the last 100 years most of the domestic wells have been sucked dry, leaving the country to import the majority of its oil from domestically unstable and expensive sources. The importation of large amounts of fossil fuel causes the *exportation* of equally large numbers of dollars that could have been invested in a North American clean energy industry.

$mart Power was written to help you, the urban homeowner or renter demystify the technology and economics required to make your journey down the road to energy efficiency and renewable energy production an enjoyable and profitable affair.

There are numerous books on ways of saving your energy dollar. Do this, change that—a hundred steps that are supposed to conserve energy and save money. The problem is that many texts do not cover the steps in an order that ensures that you are completing these upgrades in a manner that is *profitable*, *sustainable*, and *environmentally friendly*.

Family budgets are straining from the cost of energy to power our homes and fuel our cars. Many people are surprised to learn that they must work an average of two months per year just to pay their home energy bills. With the cost of all types of energy steadily increasing, the outlook for the future can only get worse.

What about the environment? Climate change is one of the key challenges facing worldwide sustainable development. Global warming, caused mainly by the burning of fossil fuels, deforestation, and inefficient use of energy, is pushing ecosystems to the brink of catastrophic failure.

Atmospheric concentrations of carbon dioxide (CO_2) have increased dramatically since the industrialization of society over the last 150 years and are at the highest levels measured during the past 500,000 years. Global energy use amongst industrialized countries is continuing to increase unabated. Underdeveloped countries, where more than one–third of the world's population does not have access to electricity, are starting to ask for their fair share of the energy pie. Exponential human population growth and accompanying energy consumption are expected to cause global temperatures to increase between 2 and 5.8°C over the coming century, according to the *Intergovernmental Panel on Climate Change* report issued in 2001.[1]

This temperature increase sounds innocent enough until you realize that a change of this magnitude has never previously occurred and will amplify weather effects throughout the earth's ecosystems. Increased levels of drought, flooding, and storms in areas already sensitized to these environmental stresses will result. The press and respected scientific papers are full of articles describing the effects of climate change: melting glaciers, reduced arctic pack ice, increasing ocean water levels, and the destruction of coral reefs. These ecologically sensitive issues are harbingers of far more devastating events to come.

So what are humans to do? The key to reversing climate change is action. The key to saving money is action. *$mart Power* will introduce you to the concept of *eco-nomics*, following a logical sequence of planning and actions that will ensure that you save money on your home energy costs while at the same time preventing damage to the world's ecology. Hence the *economic* path of energy efficiency and clean energy sources.

Energy Efficiency

North Americans need a new strategy, one that focuses on reducing energy use through efficiency and conservation rather than on increasing supply. Experience from across North America has proven that it is significantly cheaper to invest in energy efficiency than to build or even maintain polluting sources of electricity supply. This theory holds true whether we are describing the nuclear power plant down the road or your own home-based, renewable energy system. Or, as Amory Lovins, the energy guru from the Rocky Mountain Institute, is fond of saying, "It is far less expensive and environmentally more responsible to generate *negawatts* than megawatts."

Using less energy isn't about making drastic lifestyle changes or sacrifices. Conservation and efficiency measures can be as simple as improving insulation standards for new buildings, replacing incandescent light bulbs with compact fluorescent models, or replacing an old refrigerator with a more efficient one. In fact, energy efficiency often provides an improvement in lifestyle. A poorly insulated or drafty house may be impossible to keep comfortable no matter how much energy (and money) you use trying to keep warm.

California learned firsthand that saving energy means saving money and the environment. You may recall the rolling blackouts and severe power shortages that afflicted the state a few years ago. It was predicted that dozens of generating stations would be required on an urgent basis to solve the state's energy problems. At the time of writing, the total number of power stations built to solve the problem stands at zero. Faced with the realization that construction cycles for significant generating capacity would require several years, forward-thinking officials looked to energy efficiency instead. The state's energy–efficiency standards for appliances and buildings have helped Californians save more than $15.8 billion in electricity and natural gas costs. One–third of Californians cut their electricity use by 20% to qualify for a 20% rebate on their bill. The government introduced a renewable-energy buy-down and accompanying net-metering program that has seen thousands of clean, photovoltaic power systems installed on residential rooftops.[2]

In addition to saving electricity and reducing fossil fuel burning, California's conservation and efficiency efforts reduced greenhouse gas emissions by close to 8 million tonnes and nitrogen oxide emissions by 2,700 tonnes during 2001 and 2002.[2]

Not sure if you can make a significant difference at home? Consider the following points:

- Switching to compact fluorescent lamps will reduce lighting energy consumption by 80%.
- High efficiency appliances such as refrigerators, washing machines, and dishwashers can reduce energy consumption by a factor of 5 times.

- Using on-demand water heaters will reduce hot water energy costs by up to 50%.
- Low-flow showerheads, aerator faucets, and similar fittings will reduce water consumption and resulting heating costs by approximately one-half.
- A well-sealed and insulated house can reduce home heating **and** cooling costs by 50%.
- Adding a solar thermal water-heating system can further reduce hot water heating costs by 50%.

Many of the items on this list are neither expensive nor difficult to implement. Best of all, implementing these products not only dramatically reduces smog and greenhouse gas emissions, but also provides a rapid payback by inflation-proofing your home against rising energy costs and putting dollars in your wallet.

Clean Energy Sources

Sheikh Zaki Yamani, a former Saudi Arabian oil minister, sums up this position very well: "The Stone Age did not end for lack of stone, and the Oil Age will end long before the world runs out of oil." North Americans are addicted to oil in a manner that is not comprehensible to Europeans or to the developing world. While the rest of the world has developed pricing signals to achieve greater fuel and energy economy, we in North America still show up at the altar of the mighty "SUV".

The United States has less than 5% of the world's population (300 million out of 6.4 billion) and consumes a whopping 30% of *all* of the world's resources. This includes not only energy but also water, steel, aluminum, timber, and just about everything else you can imagine. When we enter a Wal-Mart or local grocery store, it is difficult to imagine that there could possibly be shortages of anything we desire. Unfortunately, present levels of consumption are not sustainable in the long term. Consider for a moment that if all of the world's population were to consume sources at the same level as the United States, we would require approximately 20 times current levels. In other words, we would require an additional seven planets worth of stuff and energy.

Indeed, with China, Russia, India and many other emerging-economy nations with combined populations many times that of the United States, it will not take long before our resources dry up and ecosystems give up in frustration.

Many people simply do not believe this. I have personally heard educated people say that climate change, air pollution, and fossil fuel shortages are the fabrications of doomsayers and quacks. However, scientific evidence suggests that these environmental hazards are not a fabrication, but pose a definite threat. Although frugality or erring on the side of caution is currently not in vogue in North America, I would suggest that this generation has a responsibility to our children and our children's children to not squander our resources.

Our current way of life includes the belief that cheap energy is our God-given right. Never mind that the cost of gasoline or imported heating oil does not include the vast subsidies lavished on the oil industry. The price of a gallon of gasoline neglects the ongoing American military presence in the Middle East, depletion subsidies, cheap access to government land, as well as monies to advance drilling and exploration technologies. None of these "hidden" costs even touches on the environmental and health damage caused by the burning of fossil fuels. For example, air pollution, largely from the burning of fossil fuels, kills an estimated 16,000 Canadians prematurely each year and results in hundreds of thousands of incidents of illness, absenteeism and asthma attacks—costing the economy billions of dollars. [3]

If the same level of direct subsidization were to be heaped on the renewable energy industry, the smoke stack would become as common as the manual typewriter.

Fossil fuels don't just power our cars. Home heating and coal-fired electrical power plants rely on fossil fuels as well. Contrary to popular belief, it is the electrical power-generation industry and not the transportation sector that contributes the largest amounts of smog, acid rain, and green house gas emissions to the world's ecosystems. [4] With much of the developing world's population connecting to the electrical grid every day, it is obvious that world energy demand will mushroom as more (and larger) appliances are brought online, further exacerbating the problem. In the developed world, middle class families are demanding more and more appliances that were considered luxury items only one generation ago. Central air conditioning, multiple refrigerators, computers, chest freezers, hot tubs, and swimming pools, all luxury or unimaginable items in our parents' day, all consume enormous amounts of energy.

Energy efficiency and demand-side management of these electrical loads will only go so far. No matter how efficient our appliances are, energy is still required to power our homes and factories. Reducing our reliance on fossil fuels and working towards cleaner energy supplies based on the efficient use of natural gas, biofuels, geothermal and renewable sources will go a long way toward creating a carbon-neutral energy supply.

The World Wildlife Fund (WWF) has introduced a program to get major power utilities to switch from high-carbon fuels—especially coal—to cleaner options. The key technologies and policies of their Powerswitch! program include:

- energy efficiency and demand-side management strategies
- large-scale wind energy projects—mainly offshore
- large-scale biomass and biomass cofiring at existing coal power stations
- increased support for renewable energies
- high rates of aluminum recycling
- rapid growth of combined heat and power systems (CHP)
- fuel switch from coal to natural gas as a transition fuel

On the home scale, several of these options are completely viable, as you will learn in this book.

By adopting demand-side management and efficient appliances, focusing on conservation, and shifting to cleaner fuels and renewable energy supplies, homeowners like you and me will be less vulnerable to energy price increases and security of supply problems and will help protect the world's ecosystems. After all, if we don't get involved, who will?

William H.Kemp
June, 2004

Footnotes
1. *Climate Change 2001 – Synthesis Report*. IPCC.
2. *Bright Future*. David Suzuki Foundation, September 2003
3. *Pembina Institute*, Autumn 2003
4. *New Scientist*, June 2003

1.2 Renewable Energy *Eco-nomics*

Now that you are motivated to "do your bit" we should take a minute to chart a path and make sure that the "bit" you "do" is the right one.

At this point you may be eager to run out and purchase a wind turbine or install solar thermal hot water heating. Not so fast! No matter how much of a "keener" you are for eco-stuff, there is right and wrong way to do everything. In fact, anyone who has had a bad experience with renewable energy systems probably didn't take the time to fully understand the technology and the economics of that decision. That is, they did not understand the *eco-nomics*.

There are many reasons for adopting renewable energy and energy efficiency systems. Too many people assume that what is right for one person is equally good for them. Alternatively, others feel that it is too expensive or only suitable for country people or for those building a new home. This is not the case.

Take a look at Figure 1-1. This graph is one of the most important pieces of information in this book. It relates the relative cost of a given technology or piece of equipment to its payback time in years as well as its financial return on investment. On the vertical or "Y" axis is increasing cost. On the horizontal or "X" axis is the approximate payback time in years.

Spread throughout the graph are various technologies related to energy efficiency and renewable energy production that can be adopted in your household. Looking more closely at the graph, you will see four distinct categories of technology:

- Basic Energy Efficiency
- Advanced Energy Efficiency
- Solar Thermal Technology
- Grid-Interconnected P.V./Wind Systems

Missing from the graph is a discussion of off-the-electrical-grid systems and transportation technologies. For those of you trying to cut the wires to your electrical utility while trying to follow an eco-nomic path, the answer is simple: don't. Although many people are sorely tempted to make the transition to a fully energy-independent home, disconnecting from the grid will never make financial sense. Besides, you don't want to lose the opportunity to sell your renewably generated electricity back to the electrical utility at a profit!

Transportation technologies have a wide array of input variables, making generalizations extremely difficult. Chapters 5 and 6 will deal with these issues in more detail.

In order to keep the graph data accurate for every jurisdiction in North

$mart Power

Renewable Energy

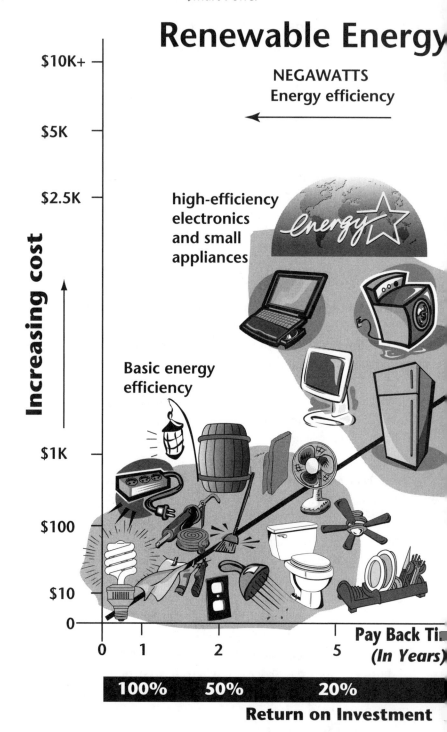

NEGAWATTS
Energy efficiency

$10K+

$5K

$2.5K

high-efficiency
electronics
and small
appliances

Increasing cost

Basic energy
efficiency

$1K

$100

$10

0

Pay Back Ti
(In Years)

0   1   2   5

100%   50%   20%

Return on Investment

ay Back Time

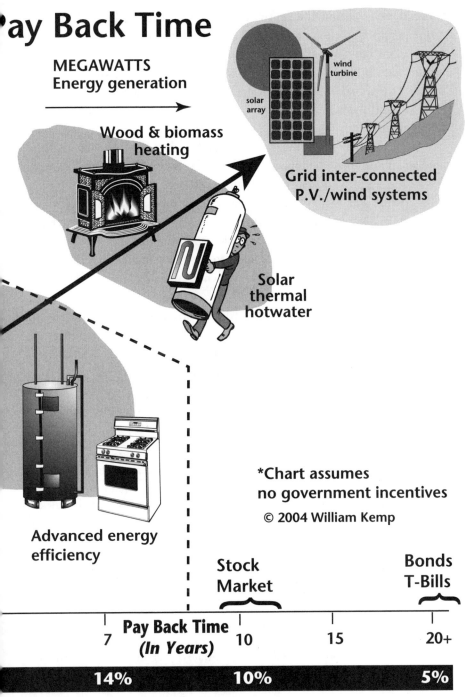

MEGAWATTS
Energy generation

Wood & biomass heating

solar array

wind turbine

Grid inter-connected P.V./wind systems

Solar thermal hotwater

Advanced energy efficiency

*Chart assumes no government incentives

© 2004 William Kemp

Stock Market

Bonds T-Bills

Pay Back Time
(In Years)

7   10   15   20+

14%   10%   5%

Figure 1-1. This graph relates the relative cost of different energy efficiency and generation technologies to payback time, without any government incentives applied.

America, no government incentive programs have been included. Obviously, if one jurisdiction is offering a 50% rebate for a given technology, this will greatly skew the payback data. The only thing you have to keep in mind as you review the payback period is that government incentives will reduce payback time, making that technology even more profitable.

Consider the issue of payback time and capital cost for any of the technologies we will be discussing. North Americans have an unfortunate habit of looking at "first cost" rather than "life cycle" or "true cost" when analyzing appliances and consumer goods. The everyday incandescent light bulb is an excellent example. Thomas Edison's mid-1800s invention was a technological marvel in its day. Replacing smoky, dim kerosene lamps with clean and bright electric lights can never be undervalued. However, just as we no longer use a horse and buggy or wireless radio sets from that era, the inefficient Edison lamp should now be relegated to the museum. The replacement is the modern compact fluorescent lamp which uses 4 1/2 times less energy and lasts 10 times longer.

At first glance, the first cost for the compact fluorescent lamp would appear uneconomic compared with the cost of the old-fashioned incandescent lamp. Typical compact fluorescent lamps have a first cost of approximately three dollars where the incandescent lamp is approximately 40 cents. But look closer. The compact fluorescent lamp lasts approximately 10 times longer, making first cost lower ($.40 x 10 incandescent lamps = $4.00, giving the same life as 1 compact fluorescent = $3.00). Based on first cost alone, our compact fluorescent lamp is already one dollar less expensive.

---

[5]   Electrical utilities are quick to point out that the cost of generation is "x" cents per kilowatt hour. What they forget to tell you is that the electricity has to be delivered to your door using transmission and distribution systems that must be paid for. Of course we can't forget the bit of tax added for the government. On average the cost of electricity delivered to your home is approximately double the cost of generation in most jurisdictions. To calculate your actual cost of electrical energy, divide your total electrical bill charges by the number of kilowatt hours of electricity used during that billing period.

[6]   At 12 cents per kWh a 100 W bulb requires $120 in electricity, a 23 W compact florescent (CF) bulb requires $27.60, saving you $92.40 in energy costs over the life of the 10,000 hour life of the CF bulb. Bulbs of varying wattages use more or less energy directly based on their power consumption compared to this example.
Incandescent Bulb = $0.12/kWh of electricity x 10,000 hours x 100 W/bulb = $120.00
CF Bulb = $0.12/kWh of electricity x 10,000 hours x 23 W/bulb = $27.60
Savings per bulb = $120.00 per Incandescent Bulb - $27.60 per CF bulb = $92.40

Now let's look at energy costs. Assuming you pay $0.12 for each kilowatt hour of electricity delivered to your home[5], a 23 W compact fluorescent lamp (which provides light equivalent to a 100 W incandescent lamp) will save over $92 in energy costs over the life of the bulb[6]. Now, multiply this savings times the average 25 light bulbs per house and you have just put a cool $2,300.00 *after-tax* dollars in your pocket.

This is real money. It's your choice: write a check to your electricity supplier or take yourself and a friend out for dinner 25 times. It's your money; why not keep the money in your pocket instead of making the utility company rich?

There is no refuting economics in this case. Businesses, cruise ships, hotels, and our European neighbors have used fluorescent lighting for years without any degradation of their lifestyle. Indeed, compact fluorescent lighting is flicker free, dimmable, of equivalent brightness and quality to incandescent lights, available in a wide selection of sizes, usable outdoors, and available with a seven-year money-back guarantee.

In addition to providing you with 20 free restaurant meals, our single 23 W compact fluorescent light will reduce greenhouse gas emissions by 1140 pounds (518 kg) and acid rain-producing compounds by 8 pounds (3.6 kg) over the life of the bulb. For the entire house, a reduction of 28,500 pounds (12,950 kg) of greenhouse gas emissions and 200 pounds (91 kg) of acid rain-producing compounds will result. As an added bonus, there will be nine fewer dead light bulbs added to your local landfill.

In this example, true cost is many times lower than first cost, providing a rapid return on investment far greater than any stock market analyst could ever (legally) achieve. The return on initial investment is only half the story. Once your initial capital is paid back, you will make a *profit equal to the return on investment*. Not even Microsoft stock in its heyday could provide the returns of the compact fluorescent light.

If you have a few hundred bucks to invest, your first reaction might be to purchase some treasury bills or bonds. However, if the bonds are giving you a 3% return and your compact fluorescent lights give you a 25% return on investment, the lamps would be the better choice.

Even if you have to borrow money from the bank to improve your home's energy efficiency, you may still be further ahead. This is particularly true for people planning on building a new home where first cost and life cycle costs of various appliances can be compared to determine the return on investment.

For example, a heat pump that has a higher first cost will almost certainly have a lower life-cycle cost than a comparable oil furnace and central air conditioning combination. Factor in rising fossil fuel costs over the same time period and the fact that the capital base of your home has increased, the investment becomes better still.

This is the case for *eco-nomics*.

It does not matter whether you own your home, rent a small apartment downtown, or are planning to build in a few years' time; *$mart Power* will guide you through the maze of energy technologies following the principles of best return on investment and environmental quality—taking the *economic* path.

Where to Start?

The vast majority of homeowners are connected to the North American electrical grid. Based on this simple statement, approximately 99.999% of the population starts the journey at the same place: the path of Energy Efficiency.

It stands to reason that any technology that is highly efficient (i.e. has a low life cycle or true cost) will have a rapid payback compared with the traditional technology it replaces. Economics and common sense tell us to begin our quest with energy efficiency at the lower left corner of the graph shown in Figure 1-1. Although the items shown in the **Basic Energy Efficiency** area of the graph are not very exciting, they do make economists' and accountants' hearts race. Why? Simple. These items are available technologies with excellent energy efficiency and rapid financial payback. I am sure all accountants have their homes equipped with compact fluorescent lamps which, as we have already learned, save 4 1/2 times the energy and operating cost of standard incandescent lamps. (If your heart can't take much more excitement, race ahead to Chapter 2 to get started).

It is interesting that all of the items listed in the **Basic Energy Efficiency** area of Figure 1 -1 apply to all people, regardless of whether they own or rent.

Once you have installed the technologies in the basic section, you can jump up a level, moving to the **Advanced Energy Efficiency** grouping. Major home appliances, including dishwashers, refrigerators, washing machines, and on-demand water heaters, appear in this area. We can also include many home electronic and electrical appliances such as television sets, stereos, and computer equipment.

Although these items are more expensive and require a longer period of time to pay back their acquisition cost, the ongoing savings are well worth the expense. Consider that a new kitchen refrigerator saves approximately $265.00 per year (see Chapter 2.4). Over a 15-year operating life, you could purchase 4 or 5 new refrigerators or go on a holiday to Alaska just on the energy savings alone, and this does not factor in rising energy costs!

Moving right along, we come to a point where we cannot squeeze any more efficiency out of our energy budget. This is the transition line between energy efficiency and energy generation. You are now ready to start producing some of your home's energy requirement as we move to the **Solar Thermal** area of the chart.

Solar thermal systems are defined as renewable energy technologies that provide space and domestic water heating. This area also includes pool and spa heating systems. You may be surprised to learn that high-efficiency wood and biomass (wood pellet) stoves are included in this category. As you will read in Chapter 3, wood and biomass energy sources are considered "carbon neutral" fuels that do not materially impact greenhouse gas emissions.

The sun's energy is very intense, providing approximately 1,000 watts of energy per square yard of area (approximately equal to 1 square meter). Up to 75% of this energy can be captured directly as heat and transferred to our domestic hot water heating needs, including potable water, swimming pools, and hot tubs.

Space heating requirements, followed very quickly by water heating, are the two largest energy consumers in the home. Home heating is also a large contributor to greenhouse gas emissions.

Solar thermal systems capture a high percentage of the sun's energy. Due to the relatively low-technology components required to capture this energy, solar thermal systems compete financially with utility-supplied electricity in many jurisdictions. For these reasons, solar thermal systems should be considered before solar electric systems, which are next on our list.

Grid-Interconnected P.V./Wind Systems have the highest "gee whiz" factor of our renewable energy technologies but are last on the payback cycle. Such is life. The reason for this is that the system's components are technically complex and difficult to manufacture. (We will touch on these items in Chapters 8-12).

Many jurisdictions recognize the numerous advantages of distributed generation or grid-interconnected electrical generation. Consider that photovoltaic (PV) electric systems produce maximum power and support to the electrical grid during times of peak demand (daylight hours). If distributed systems were installed in sufficient numbers, expensive electrical generation "peak demand" plants would not have to be built. As if that weren't reason

Figure 1-2. Many jurisdictions recognize the value of "distributed energy generation" by offering generous incentives to produce your own electricity and sell any surplus to the electrical utility. This grid-interconnected urban home is an example of the future being here now. (Courtesy Sharp Electronic Corporation)

enough to consider the technology, local homeowners are paid for this valuable energy through net-metering programs, injecting more money into the local economy.

In further recognition of the value of these systems, many jurisdictions are offering generous incentives to "pay down" the up front capital cost. In addition, time-of-day metering is often permitted, thereby paying home-generators a premium for their electrical generation which is coincident with periods of peak demand.

As of the time of writing, California has one of the best buy-down programs, wherein up to 50% of the installed cost of the system is credited back to the homeowner. Many states and provinces are following California's lead. (A database of state incentives for renewable energy is available at the DSIRE website, www.dsireusa.org).

What about Off–the–Grid Systems?
As discussed above, off-the-grid electrical systems are simply not economic in an urban environment. You may be tempted to consider an off-grid system to reduce your $100 per month electrical bill. If you've been following along so far you will understand that it is far more economical to reduce consumption than to generate more energy. Therefore, the first step would be to control consumption in order to decrease the $100 electrical bill considerably. By the time you've reduced your electrical bill, a $20,000 off-grid system that is required to produce $20 worth of electricity won't seem like such a good idea after all. Using the same logic, a $100-per-month electrical bill would require an off-grid system costing at least $100,000 to generate

such a large amount of energy.

"Off–the-grid" or simply "off-grid" systems have a unique way of looking at payback criteria. In most (but certainly not all) cases, a homeowner who wishes to build a home or cottage on a desirable lot which is a distance away from the utility lines must pay for the line extension to the home. One does not have to run the wire very far before a serious amount of cash is burned up. It may only require a line extension of 0.5 miles (0.8 km) to exceed the cost of an off-grid system. In that case, the payback time is the instant the first light is turned on!

Obviously, the further away from the utility lines, the faster the payback, to the point where an off-grid system may be the only economic way to build in that location. Islands and mountaintops come to mind.

"Off gridders" (that group of people who are often fond of granola and Birkenstocks) can use the cost of utility-line extensions to their advantage. A building lot that has no access to electrical power might just not be worth as much in the seller's eyes, even though the buyer has no intention of making this connection.

As an urban dweller, stick with grid-interconnected systems.

Eco-nomic Conclusion

Now that we have a bit of insight into the *eco-nomic* reasons for doing things, I must ask you a simple question: What is the payback on your boat, your Lexus, or your big screen TV? The answer of course is never. So why did you purchase any of these items in the first place? Choice. Simple market-driven choice. This explains why accountants and economists are such a boring bunch. They never buy luxury goods unless there is a sound economic reason for the purchase.

Choice is a luxury we are blessed with in North America. Discretionary income is very high and (except for accountants and economists) we can purchase items based simply on choice. The reason for your purchases is based on your own personal goals, needs, and wants. Renewable energy technologies are no exception. If helping the planet, being an early adopter, or bragging to the neighbors is your thing, then throw out Figure 1-1 and move on to Chapter 2.

> **A Word about Pricing:** Many of the products described in this handbook are either manufactured in the United States or priced in U.S. dollars. To avoid confusion, all prices are shown in U.S. dollars.

1.3 What Is Energy?

All of the earth's energy comes from the sun. In the case of renewable energy sources and how we harness that solar energy, the link is often very clear: sunlight shining through a window or on a solar heating panel creates warmth, and when it strikes a photovoltaic (or PV) panel the sunlight is converted directly into electricity; the sun's energy causes the winds to blow, which moves the blades of a wind turbine, causing a generator shaft to spin and produce electricity; the sun evaporates water and forms the clouds in the sky from which the water, in the form of raindrops, falls back to earth becomes a stream that runs downhill into a micro-hydroelectric generator.

While these energy sources are renewable, they are also variable and intermittent. The sun generally goes down at night and may not shine for several cloudy days. If we just wanted to use our energy when it is available, the various systems used to collect and distribute it would be a whole lot simpler. But we humans are just not that easily contented. I know for a fact that most people want their lights to turn on at night, even though the sun stopped shining on the PV panels hours ago. In order to ensure that heat and electricity are available when we need it, a series of cables, fuses, boilers, and a bewildering array of components are required to capture nature's energy and deliver it to us on demand.

The process of capturing and using renewable energy may seem far too complicated and expensive for the average person. However, while the components themselves are complicated, the theory and techniques required to understand, install, and live with renewable energy are not.

Figure 1-3. Most people want their lights to turn on at night. This little issue complicates the requirements of a renewable energy system.

Why Are We Discussing Math?

Although it is not absolutely necessary to be an expert in heating, wind, and electrical energy, an understanding of the basics will greatly assist you in operating your renewable energy system. This knowledge will also help you make better decisions when it comes to purchasing the most energy-efficient appliances or figuring out your previously indecipherable hydro bills.

If you are planning a grid-interconnected system, you may consume as much energy as you like, provided you don't mind paying for it. For off-grid systems, energy consumption is an important consideration, as additional energy comes from more expensive and complex alternative sources or fossil-fuel generators.

Trying to save money by reducing energy costs can only be accomplished by understanding your "energy miles per gallon" quotient. It is fairly simple to determine gas fuel economy or efficiency, but it is much more difficult to determine energy efficiency for a house full of appliances and heating equipment.

The mathematical theories described in *$mart Power* are simplified. Equations are used throughout the text, but they are limited to those areas that are most important. While the mathematics might be a bit light for some, for the rest of us the math and logic are fairly straightforward; a simple calculator (powered by the sun of course) will make the task that much simpler. For those so inclined, there are plenty of references to data sources that will make even the most ardent "techie" happy.

The Story of Electrons

If you can remember back to your high school days in science class, you may recall that an atom consists of a number of electrons swirling around a nucleus. When an atom has either an excess or a lack of electrons in comparison to its "normal" state, it is negatively or positively charged, respectively.

In the same way that the north and south poles of two magnets are attracted, two oppositely charged atoms are also at-

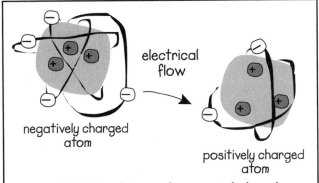

negatively charged atom

electrical flow

positively charged atom

Figure 1-4. The flow of electrons from negatively charged atoms to positively charged ones is known as the flow of electricity.

tracted. When a negatively charged atom collides with a positively charged one, the excess electrons in the negatively charged atom flow into the positively charged atom. This phenomenon, when it occurs in far larger quantities of atoms, is called the flow of electricity.

The force that causes electricity to flow is commonly known as *voltage* (or V for short). The actual flow of the electrons is referred to as the *current*. So where does this force come from? What makes the electrons flow in the first place? The trigger that brings about the flow of electrons can come from several energy sources. Typical sources are chemical batteries, photovoltaic cells, wind turbines, electric generators, and the up-and-coming fuel cell. Each source uses a different means to trigger the flow of electrons. Waterfalls, coal, oil, or nuclear energy are commonly used to generate commercial electricity. Fossil or nuclear fuels are used to boil water, which creates the steam that drives a turbine and generator; falling water drives a turbine and generator directly. The spinning generator shaft induces magnetic fields into the generator windings, forcing electrons to flow.

Let's use the flow of water as a visual aid to understanding the flow of electricity that is otherwise invisible.

Let's presume that a greater amount of water moving past you per second equates to a higher flow and that a river with a high flow of water has a large current. With electricity, a large number of electrons flowing from one atom to the next is similar to a large flow of water. Therefore, the greater the number of elec-

small creek

high waterfall

Figure 1-5. The flow of electricity is very similar to the flow of water. A waterfall has a higher flow and greater pressure than a creek; similarly, a higher number of electrons moving from atom to atom increases electrical current.

trons flowing past a given point per second, the greater the electrical current.

The flow of water is typically measured in gallons per minute or liters per second; electrical current is measured in amperes (or A for short). If the measured current of electrons in a material is said to be 2 A, we know that a certain number of electrons have passed a given point per second. If the current were increased to 4 A, there would be twice the number of electrons flowing through the material.

Gravity is a factor in considering water pressure. A meandering creek has very little water pressure, owing to the small hill or height that the water has to fall. A high waterfall has a much greater distance to fall and therefore has increased water pressure. For example, if a water-filled balloon were to fall on you from a height of 2 feet you would find this quite refreshing on a hot day. A second balloon falling from a height of 100 feet would exert a much higher pressure and probably knock you out! This increase in water pressure is similar to the electrical pressure or voltage that forces the electrons to flow through a material.

Conductors and Insulators

Electricity flows through conductors in the same manner as water flows through pipes and fittings. When electrons are flowing freely through a material we know that the material is offering little resistance to that flow. These materials are known as conductors. Typical conductors are copper and aluminum, which are the substances used to make electrical wires of varying diameters. Just as a fire hose carries more water than a garden hose, a large electrical wire carries more electrons than a thin one. A larger wire can accept a greater flow of current (i.e. handle higher amperage).

Before electricity can be put to use, we must create an electrical circuit. We do this by connecting a source of electrical voltage to a conductor and causing electrons to flow through a load and back to the source of voltage.

The drawing in Figure 1-6 shows how a simple flashlight works. Electrons stored in the battery (we will cover that one later) are forced by the voltage (pressure) to flow into the conductor from the battery's negative contact (-) through the

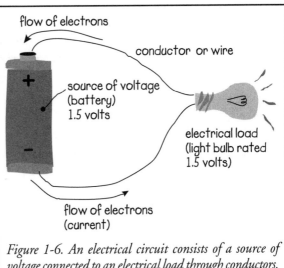

flow of electrons

conductor or wire

source of voltage (battery) 1.5 volts

electrical load (light bulb rated 1.5 volts)

flow of electrons (current)

Figure 1-6. An electrical circuit consists of a source of voltage connected to an electrical load through conductors.

light bulb and back to the battery's positive contact (+). Current flowing in this manner is called *direct current* (or DC for short). The light will stay lit until the battery dies (runs out of electrons) or until we turn off the switch.

Oh yes, a switch would be a good idea. This handy little device allows us to turn off our flashlight to prevent using all of the electrons in the battery. From our description of a circuit, we can assume that if the electrical conductor path is broken, the light will go out. How do we break the path? The flow of electrons may be interrupted using a nonconductive substance wired in *series* with the conductor. Any substance that does not conduct electricity is known as an insulator.

Typical insulators include rubber, plastics, air, ceramics, and glass.

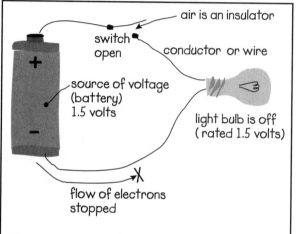

Figure 1-7. Any substance that does not conduct electricity is an insulator.

Batteries, Cells, and Voltage

You may have noticed that your brand of flashlight has two or even three cells. Placing cells in a stack or in *series* causes the voltage to increase. For example, the flashlight shown in Figure 1-7 contains a cell rated at 1.5 V. Placing two cells in series, as shown in Figure 1-8, increases the voltage to 3 V (1.5 V + 1.5 V = 3 V). The light bulb in Figure 1-8 is glowing very brightly and will quickly burn out because it is rated for 1.5 V. In this example, a higher voltage (pressure) is causing more current to flow through the circuit than the bulb is able to withstand. Likewise, if the cell voltage were lower than the rating of the bulb, insufficient current would flow and the bulb would be dim. This is what happens when your flashlight batteries are nearly dead and the light is becoming dim: the batteries are running out of electrons.

Next time you are waiting in the grocery store lineup, put down that copy of *The National Enquirer* and take a look at the battery display. The selection will include AA, C, and D sizes of cell, which all have a rating of 1.5 V. Can you guess the difference between them?

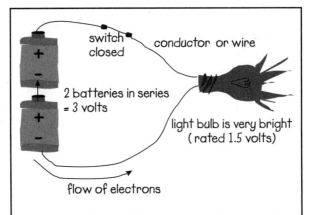

A larger cell holds more electrons than a smaller one. With more electrons, a larger battery cell can power an electrical load longer than a smaller one. The circuit shown in Figure

Figure 1-8. A higher voltage forces more electrons to flow (higher current) than the bulb is rated to tolerate. Always ensure that the source and load voltage are rated equally.

1-10 shows a set of jumper wires connecting the cell terminals in *parallel*. This parallel arrangement creates a battery bank of 4 AA-size cells, which has the same number of electrons as and the capacity of the C-size cell. Any grouping of cells, whether connected in series, in parallel, or both, is called a battery bank or battery.

Obviously, a house requires far more electricity than a simple flashlight does. Off-grid homes generate electricity from renewable sources (more on that later) and may store it in battery banks such as the one shown in Figure 1-11.

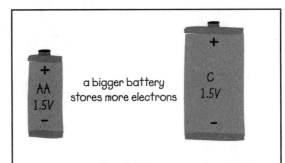

Each battery cell is numbered 1 through 12 and has a nominal voltage of 2 V. If you look carefully, you can see that each cell has a (+) and (-) terminal. Each terminal is wired in *series* to the next battery in the manner illustrated in Figure 1-8.

Figure 1-9. Cells of different sizes have different amounts of electron storage. Just as a 2-liter jug contains twice as much water as a 1-liter bottle, so a larger battery stores more electrons and works for longer periods than a smaller one.

Therefore a series string of 12 cells rated at 2 V each creates a battery bank rated at 24 V DC (Vdc).

You will also note that there are two such banks of batteries. The bank on the left is identical to the bank on the right. By wiring the two banks in *parallel* we create a total battery capacity that is twice as large as a single bank. This is exactly the same as wiring the 4 AA cells in parallel to make the equiva-

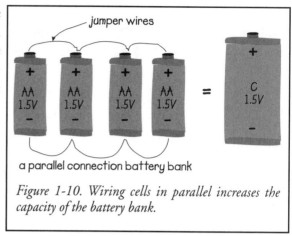

a parallel connection battery bank

Figure 1-10. Wiring cells in parallel increases the capacity of the battery bank.

lent of the large C cell. Obviously, the more electricity we use the greater the battery size and cost.

Figure 1-11. Off-grid homes or homes equipped with emergency backup systems require large amounts of electricity, which is stored in a battery bank such as this one.

Grid-Connected Houses Don't Run on Batteries

You are probably aware that your house works on 120/240 V and not 12 V, 24 V, or 48 V from a battery. Early off-grid houses, small cottages, boats, and many recreational vehicles can and do use 12 Vdc systems. But don't consider using low-voltage DC for anything but the smallest of systems. With the limited selection of appliances and the difficulty of wiring a low-voltage, full-time home, low-voltage DC systems are not a viable option for homes.

The modern grid-connected home is supplied with electricity by an electrical utility in the form of 120/240 V *alternating current* (Vac). In a DC circuit, as defined earlier, current flows from the negative terminal of the battery through the load and back to the positive terminal of the battery. Since the current is always flowing one way, in a direct route, this is called direct current.

In an AC circuit, the current flow starts at a first terminal and flows through the load to the second terminal, similar to the flow in a battery circuit. However, a fraction of a second later the current stops flowing and then reverses direction, flowing from the second terminal through the load and back to the first terminal, as shown in Figure 1-12.

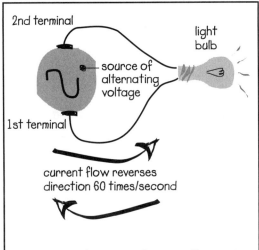

Figure 1-12. The process of continually reversing the direction of current flow is known as alternating current.

Alternating Current in the Home

Generating electricity in the modern, grid-connected world is accomplished by using various mechanical turbines to turn an electrical generator. In the early years of the electrical system, there was considerable debate as to whether the generator output should be transmitted as AC or DC. For a time, both AC and DC were generated and transmitted throughout a city. However, as time passed, it became clear for safety, transmission, and other practical reasons that AC was more desirable. As the old saying goes, "The rest is history." All modern houses and electrical appliances are standardized in North America to operate on either 120 Vac or 240 Vac. For this reason, it is advisable for off-grid homes to convert the low-voltage electrical energy stored in the battery bank to 120/240 Vac. Likewise, for renewable energy systems that are grid interconnected, it is necessary to convert the DC voltage from the photocells or wind turbine to 120 Vac or 240 Vac for sale and storage on the grid. The device used for this electrical conversion is the inverter, which we will discuss in more detail in Chapter 12.

Power, Energy, and Conservation

Conserving energy and doing more with less is not only good for the planet but also helps keep the size and cost of your renewable-energy power station within reasonable limits. Lowering electrical house loads also increases the amount of electricity you can *sell* to your electrical utility, increasing your power station revenue. We discussed earlier how a bigger battery could run a light bulb longer than a smaller battery. This is fairly obvious. What might not be so obvious is that if we replace the "ordinary" light bulb with a more efficient one we might not need the larger battery in the first place.

Let's review how any typical electrical circuit operates: A source of electrons under pressure (i.e. voltage) flows through a conductor to an electrical load and back to the source. For household circuits, the voltage (pressure) is

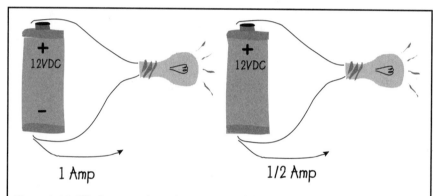

Figure 1-13. *The lamp on the right uses 50% of the electrical energy of the one on the left and both have equal brightness. The lamp on the right is said to be twice as efficient.*

usually fixed at 120 Vac or 240 Vac. For battery-supplied circuits, the voltage is usually fixed at either 12 Vdc, 24 Vdc, or 48 Vdc, depending on the size of the load and the amount of electron flow (i.e. current) that is required to make the load operate. Let's say that an ordinary light bulb requires 12 V of pressure and 1 A of current flow to make it light. Now suppose that we can find a light bulb that uses 12 V and only 0.5 A of current flow. Assuming that both lights are the same brightness, we infer that the second light is twice as efficient as the first. Stated another way, we would need only half the battery-bank size (at lower cost) to run the second lamp for the same period of time, or the same size of battery bank would run the second lamp for twice as long.

The relationship between the pressure (voltage) required to push the electrons to flow in a circuit and the number of electrons actually flowing to make the load operate (current or amps) is the power (commonly expressed as *watts*, or W) consumed by the load. Using the above example, let's compare the power of the two circuits:

Ordinary Lamp:

12 V x 1 A = 12 W

More Efficient Lamp:

12 V x 0.5 A = 6 W

The second lamp uses half the number of electrons to operate. The power of a circuit is simply the voltage multiplied by the current in amps, giving us the instantaneous flow of electrons in the circuit measured in watts.

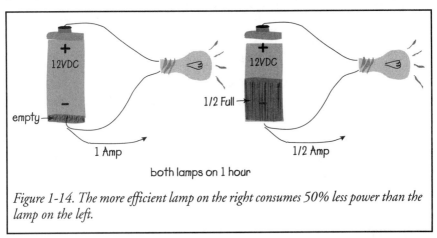

Figure 1-14. The more efficient lamp on the right consumes 50% less power than the lamp on the left.

Remember that the more efficient bulb operates twice as long on a battery as the other bulb. How does time factor into this equation? If a battery has a known number of electrons stored in it and we use them up at a given rate, the battery will become empty over time. The use of electrons (power) over a period of time is known as *energy*.

Energy is *power* multiplied by the *time* the load is turned on:

Ordinary Lamp:

12 V x 1 A x 1 hour = 12 Watt hours; or

12 W x 1 hour = 12 Watt hours

More Efficient Lamp:

12 V x 0.5 A x 1 hour = 6 Watt hours; or

6 W x 1 hour = 6 Watt hours

Your electrical utility charges you for energy, not power. You run around turning off unused lights to cut down on the amount of time the lights are left on. If we think about it long enough, we can also use these calculations to figure out how much energy is stored in a battery bank. For example, if a 12 V battery bank can run a 10 A load for 30 hours, how much energy is stored in the battery?

Battery Bank Energy (in Watt Hours):
12 V x 10 A x 30 hours = 3,600 Watt hours or 3.6 kilo-Watt hours of energy

An interesting thing about batteries is that their voltage tends to be a bit "elastic." The voltage in a battery tends to dip and rise as a function of its state of charge. For example, our nominal 12 V battery bank only registers 12 V when the batteries are nearly dead. When they are under full charge, the voltage may reach nearly 16 V. Because of this swing in the voltage, many batteries are not sized in *Watt hours* of energy, but *Amp hours*.

The math is similar; just drop the voltage from the energy calculation:

Battery Bank Energy (in Amp Hours):
10 A x 30 hours = 300 Amp hours of energy

To convert Amp hours of energy to Watt hours, simply multiply Amp hours by the battery voltage. Likewise, to convert Watt hours of battery bank capacity to Amp hours, divide Watt hours by the battery voltage.

That pretty well covers everything you need to know about electrical energy. With a bit of practice (and you'll get plenty of that in the chapters to follow) you will know this stuff well enough to brag at your next office party.

Heating Energy

If you happen to be sitting beside a nice warm wood stove as you read *$mart Power*, you can feel the heat from the fire radiating towards you. This warmth and the effects of reading the section above on electrons may have caused you to doze off to sleep. If not, you might consider that heat is not the same thing as electricity, for if it were, it could not reach you since air is an insulator.

In fact, the idea that heat is a form of energy baffled many earlier physicists and took a long time to be understood. It took an even longer period for the theory of heat to make its effect on industry. Consider that houses built around the turn of the century gave no thought to conserving heat by means of insulating or any of a number of techniques which seem quite obvious to us now.

So what exactly is heat? Early researchers thought that heat was a substance, something you could put in a bottle—a fluid they called *caloric*. This theory persisted until the middle of the 19th century and was not nearly as silly as you might think. Consider: Heating a liquid causes it to expand, as if something is added to it; when wood is burned, a small pile of ashes is left behind, as if something has escaped or evaporated; a jar of boiling hot water placed next to a jar of cool water causes the cool water to warm up. Perhaps *caloric* flowed from the warmer jar to the colder?

Although we now understand that the notion of *caloric* is incorrect, the concept of heat flowing as if it were a fluid is not that far from the truth, as the above examples illustrate. As scientists continued scratching their heads, the current laws of heat and thermodynamics slowly replaced the theory of *caloric*.

It is now understood that heat is a form of energy caused by the motion of molecules, or groups of atoms. All matter is made up of molecules, which (like

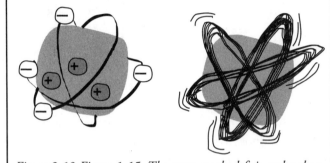

Figure 2-13 Figure 1-15. The atom on the left is cooler than the one on the right because it is moving more slowly.

teenagers) are always in motion. As a substance is warmed, its molecules move faster; as it is cooled, its molecules move less. The temperature of a substance is directly related to the motion of these tiny molecules.

One of the primary practical considerations relating to heat is that it does not like to stay still (also like teenagers). Consider any type of heat and you will notice that it always wants to move from the hotter object to a colder one. A hot pan placed in cold water will cool, while warming the water. In the same manner, in winter your expensive heated air wants to get outside as quickly as possible to help melt the snow. Although this might be a useful task, the cost in dollars and energy is just a bit beyond reach!

When we wish to stop the flow of electricity in a circuit, we use a non-conductive device called an insulator. Heat energy can also be slowed down in its relentless path into or out of our homes by the use of an insulator for heat (insulation), which is discussed in Chapter 2.

Because heat is energy, it is also possible to quantify this energy in the same way that electricity is quantified in kilowatt-hour units. As a matter of fact, if we were to heat, cool, and operate our homes completely on electricity, the energy usage charge would indeed be in kWh, as all of the energy used would be sourced from the electrical utility. However, as you know there are many sources of heating/cooling energy to choose from, including natural gas, electricity, propane, oil, wood and wood by-products, and even coal. Each of these sources of heat energy is delivered to you in a bewildering array of units, making comparison shopping very difficult. To complicate matters further, the efficiency of the heating or cooling equipment using these various sources varies greatly.

To level the playing field, we will use the English "British thermal unit," or BTU, as our standard for comparing heating and cooling energy. For readers more accustomed to the metric system, the calorie or joule measure will provide the same basis for comparison. Just make sure you don't use both systems at the same time, or you may be wondering why you have to chip ice out of your toilets next January.

One BTU is the quantity of heat required to raise, by one Fahrenheit degree, the temperature of one pound of water. Using these units allows you to quickly compare two heating sources which have different base units. For example, suppose you are trying to compare the cost of heating a house with oil or propane. Which is more economical?

Assume that the current cost to heat your home is $500.00 per year and that you require 250 gallons of oil at $2.00 per gallon. Propane costs $1.75 per gallon. You assume that propane costs 25 cents less per gallon and that it will be cheaper to use. Sorry, it doesn't work like that. The first step is to find out how much heat energy in BTUs is stored in each energy source. Refer to the cross-reference chart in Appendix 1 and look up information on both energy sources. You will see that oil contains 142,000 BTUs per gallon and propane contains 91,500 BTUs per gallon. Therefore:

250 gallons of oil x 142,000 BTU/gallon = 35,500,000 BTU per heating season is required to heat your house.

And, if we were to assume that we need 250 gallons of propane:
250 gallons of propane x 91,500 BTU/gallon = 22,875,000 BTU

But your house requires 35.5 million BTU, so we have a shortfall of:
35,500,000 required – 22,875,000 from propane = 12,625,000 BTU shortfall

Now let's take a look at the costs. The current heating charge using oil is $500.00, and our assumption that 250 gallons of propane would be sufficient produced our first estimate of:

250 gallons of propane x $1.75 per gallon = $437.50 or a savings of $62.50/year

However, in order to make up the shortfall of 12+ million BTU, we have to purchase more propane:

12,625,000 BTU shortfall ÷ 91,500 BTU/gallon propane = 138 more gallons

138 more gallons x $1.75 / gallon of propane = $241.50 additional cost

The total cost of heating your house with propane is now *$437.50 + $241.50 = $679.00.*

As you can see, your cost for using propane over heating oil went from an estimated saving of $62.50 to an increase of $179.00 per year! Of course these costs are not real, but the example shows that the amount of heat energy "stored" in a fuel is known and can be used for comparison when all fuels have the same base units of BTU or joules. This also applies to renewable sources such as firewood, wood pellets, and even solar heating systems.

You must also consider the efficiency of the heating appliance. Suppose you find two nearly identical heating devices at the same price with output ratings of 120,000 BTUs per hour. (The fuel type is not important for this calculation). One has an efficiency of 80% and the other 65%. It stands to reason that the 80%-efficiency-rated unit is the better buy. Over the lifetime of a product, this difference in efficiency can add up to tens of thousands of dollars.

Not interested in doing the math? Contact a competent home energy dealer and discuss these issues. Most dealers can show how fuel prices affect the operating cost of the heating appliances they sell. You will have to go to the fortune teller down the street to determine future energy costs. The one thing you won't need a fortune teller for is to understand that *all* energy costs are rising.

Cooling Energy

An important consideration for homeowners is how to "make" cool air in the dog days of summer. No matter how big your air conditioning unit is, you do not make cold air. A better way to understand a mechanical cooling system is to think of it as a *heat pump*. Modern air conditioning units are complex devices that move heat from one location to another. A window air conditioner pumps heat from your warm indoor room and moves it outside. This may seem a little difficult to understand until you realize that the outside part of the A/C unit is quite hot because of the indoor heat moved there by the internal mechanism. As the outside component (the condenser) is hotter than the outdoor air, the heat (guess what?) wants to move from the hot condenser to the relatively cooler outdoor air. This phenomenon of air-conditioning heat transfer explains why the unit has to work harder in hotter weather. As the outside air temperature approximates that of the condenser, heat transfer slows, making the unit run for longer periods. This is also why you want to place the condenser facing north or out of direct sunlight.

This cooling process is not efficient in even the best-designed air conditioners or more complex heat-pump systems. Regardless of the type of A/C system installed, the energy used to cool a given area is measured either by the BTUs of heat moved or by the electrical energy consumed by the unit. Commercial A/C units are usually rated both ways, with some units still carrying a "cooling tonnage" rating. This archaic rating compares the cool-

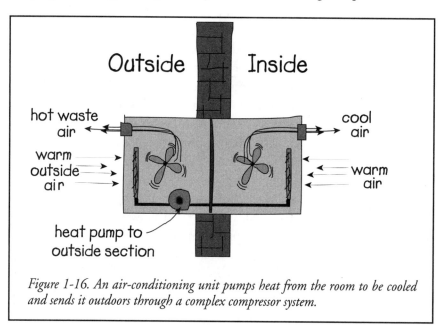

Figure 1-16. An air-conditioning unit pumps heat from the room to be cooled and sends it outdoors through a complex compressor system.

ing capacity of the A/C unit to a ton of ice. Just in case you were wondering, this is approximately equal to 12,000 BTU of cooling capacity.

With all of the heating and cooling energy flowing here and there it seems that just maybe the *caloric* theory really should have been given more consideration after all. At least we could measure the stuff in gallons or liters and not worry about BTUs and cooling tons.

1.4 Energy and Climate Change

Everyone has heard about them, but do you actually know what smog, climate change, and greenhouse gas emissions really are? Ask anyone who lived in Los Angeles in the early 1980s and they will tell you about dirty skies and endless haze. For someone with asthma or other respiratory sensitivities it meant staying indoors for countless hours or depending on medical inhalers to struggle with each breath. Steve Ovett wished the air were cleaner in Los Angeles. Although he effortlessly beat his rivals during the 800- and 1,500-meter Olympic men's races, going 3 years and 45 races undefeated, the air in Los Angeles was his undoing. During the 1984 Olympics he collapsed during the 1500 metre race due to smog-induced asthma and spent two nights in hospital.

Smog and soot pollution is created when the emissions from burning fossil fuels combine with atmospheric oxygen and ultraviolet light from the sun. Estimates indicate that over half a million deaths worldwide are directly attributed each year to excessive levels of airborne pollution. When smog and soot particles combine with raindrops, acid rain is formed.

Clean rain water is neither acidic nor base, meaning that this life-giving element is non-corrosive and restorative to everything it touches. Acid rain, on the other hand, is precisely that—water which has become acidic. When it comes into contact with metal, rock, or plants the result is a corrosive

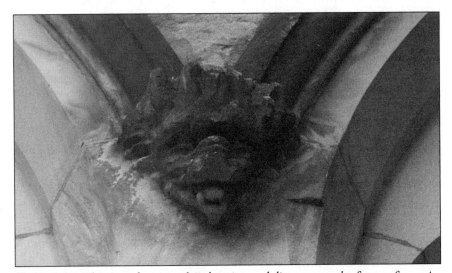

Figure 1-17. This carved stonework is decaying and dirty as a result of years of corrosive acid rain and soot accumulations. Many of the world's priceless carvings and architectural treasures are rotting away in a similar manner.

action and the decay and destruction of a wide array of living and lifeless objects.

Smog has been greatly reduced in the developed world in recent years as a result of decreased sulfur concentrations in gasoline and diesel fuels. Improvements in automotive and coal-fired power-station emissions have also gone a long way in reducing smog over the last 20 years; however, the problem is far from over. Increasing demand for coal-fired electrical energy and ever-increasing numbers of automobiles on the world's highways continue to exacerbate the problem.

Carbon dioxide and methane are two greenhouse gases that are increasing in concentration in the atmosphere and are directly linked to global warming. Carbon dioxide concentrations are higher than at any time in the last 420,000 years and have directly contributed to a rise in the Earth's surface temperature of 0.6 degrees over the last century. It should also be noted that the 1990s were the warmest decade on record, with 1998 recorded as the single warmest year of the last 1,000 years.

Scientists predict that if the earth's average temperature were to rise by approximately 2°C, numerous devastating effects could occur:

- increased magnitude and frequency of weather events
- rising ocean levels causing massive flooding and devastation of low-lying areas
- increased incidents of drought affecting food production
- rapid spread of non-native diseases

So where do these gases come from? Carbon dioxide is a byproduct of the burning of any carbon-based fuel: gasoline, wood, oil, coal, or natural gas. With few exceptions, the world's current energy economy is fueled by carbon.

We have to go back half a million years and more to learn how this carbon fuel came to be. All carbon fuel sources began as living things in prehistoric times. Peat growing in bogs absorbed carbon dioxide from the air as part of the photosynthetic process of plant life. The rolling, heaving crust of the earth entombed the dead plant material. Sealed and deprived of oxygen, the plant matter could not rot. Over the millennia, shifting soils and ground heating compressed the organic material into soft coal that we retrieve today from shallow open-pit mines. Allowed to simmer and churn longer, under higher heat and pressure, oil and hard coal located in deep underground mines are created.

Provided these prehistoric fuel sources remain trapped underground there is no net increase in atmospheric carbon dioxide. However, the process of burning any of these fuels reduces the carbon stored in the coal, oil, or wood and drives off the trapped CO_2.

It is not necessary for society to make a wholesale switch to a non-carbon-based fuel such as hydrogen in order to reduce the effects of global warming. By selecting a more fuel-efficient car, switching to compact fluorescent lamps, or improving the energy efficiency of our home we can drastically reduce greenhouse gas emissions and enjoy significant savings along the way.

From an *eco-nomic* point of view, doing more with less is the path to sustainable development.

2
Energy Efficiency

When the price of gasoline jumps for the tenth time in a month, most people tend to cut back on their driving—or at least to talk about it. Some people might even consider trading their vehicle in for something with better gas mileage than a Hummer.

Family budgets are straining from the rising cost of living: taxes, mortgages, car payments, and energy bills for homes and vehicles. As a society we rarely consider the source of these costs. People hop on the bandwagon of the middle-class dream fueled by the two incomes that make it happen. Large homes in the suburbs require enormous amounts of heat, light, and air conditioning, and many people have two cars to get to distant jobs. All of this eats away at our precious discretionary income—and our free time.

Figure 2-1 An energy-conserving lifestyle does not mean a spartan lifestyle. Choosing the most energy-efficient appliances and products that "do more with less" is how to make it happen. The hybrid 2004 Toyota Prius doubles gas mileage efficiency and reduces pollutants by 90%. Courtesy Toyota Corporation.

Doesn't energy conservation mean giving things up, not living the middle-class dream North Americans have come to expect, or living a spartan life of near poverty? Well, yes and no. Yes, energy conservation does mean giving things up, but they are mostly wasteful, inefficient things. And no, it does not mean that you must compromise the quality of your lifestyle.

How can we reduce our heating and electrical loads or transportation costs to embrace energy conservation without giving anything up? The answer is efficiency. As the current corporate downsizing mantra goes: "Do more with less."

I recently heard a politician asking us all to do our part to conserve energy by restricting ourselves to five-minute showers. This is a misdirected effort. People may do their part by conserving energy or resources in times of severe famine, war, or national disaster, but it is never sustainable. As soon as things are back to normal people immediately return to their old habits. Conservation is not sustainable if there is a perception that it will interfere with quality of life.

The better approach is to take a ten-minute shower but use a low-flow showerhead. Installing this five-dollar device will not reduce your quality of life and in fact might improve it, as many models incorporate a massage feature. By reducing energy and water consumption by 50% or more, they will put real money in your pocket.

This is the *eco-nomic* approach: doing more with less.

It may sound like nickel-and-dime stuff, but in fact the opposite is true. The constant "leakage" of energy dollars here and there can become quite significant. Refer back to Chapter 1 and you will recall that the simple act of switching common incandescent lamps to high-efficiency compact fluorescent models will put over $2,300 *after-tax* in your pocket over the operating life of the lamps. If you are in a 35% tax bracket this translates into a savings of over $3,100.

This is just one example of dozens of tips that, added together, translate into serious energy and financial savings. What you do with the savings is up to you. Have you been trying to scrape together enough cash to take the kids on vacation or top up your retirement savings plan? If so, this is one of the easiest ways to do it.

Let's start our quest for home-energy efficiency with an understanding of the house itself. First, we'll look at some of the most obvious design features that can be incorporated into the building. Next, we'll take a look around inside. Lighting, heating, electronics, and major appliances all consume energy and must be considered in an energy-efficient lifestyle.

2.1
New Home-Design Considerations

By far the largest energy usage in North American homes is heating and cooling. As we learned in Chapter 1.3, heat has a nasty habit of wanting to escape outside in winter. In summer, the opposite is true. Heat from the hot summer sun just can't wait to come inside and enjoy the air conditioning with you.

It is not necessary to design your own home in order to achieve satisfactory energy-efficiency results. Many builders now recognize that energy-efficient homes can almost compete with granite countertops and hardwood floors in attracting the buying public. Look for builders that are certified for R-2000 or other high-quality construction methods.

Heat Loss and Insulation

Heat loss and gain in your home is linked to the level and quality of insulation in the ceilings, walls, and basement. Not so obvious are the losses associated with windows, doors, and the sealing in various joints and holes in the structural cavity. The key to designing an energy-efficient house is to ensure that it is well constructed and airtight and contains adequate levels of insulation.

The national and local building codes in your area set the *minimum* standards that your building must meet, but it is fairly easy to incorporate upgrades into your design that will pay for themselves many times over. With the world's current political climate and uncertainty over fossil fuel supplies in the coming years, these upgrades will provide a safety net against increasing energy prices.

Insulation in the walls and ceilings of our homes slows the transfer of heat into or out of our house (in much the same manner that electrical energy cannot flow through a non-conductive material). The

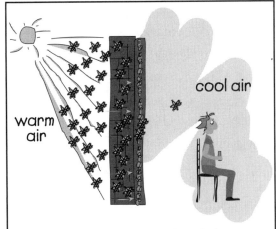

cool air

warm air

Figure 2-2. Heat loves to move from the hot summer outdoors to enjoy the air conditioning inside with you. Some materials such as brick and stone allow heat to travel fairly easily. Proper insulation slows the flow of heat.

higher the quality and thickness of the insulation, the harder time the heat has getting through. Many people believe that because hot air rises, most of the heat loss in the house will be up through the ceiling. Not so. Heat moves wherever it can, shifting from warm areas to colder ones, whether that be upwards, downwards, or sideways. Keep this in mind as we visit all areas of the home during our insulation spree.

Table 2-1 lists typical recommended insulation values for a very well insulated home. The "R" value (English system) and "RSI" value (metric system) indicate quality levels of insulation. The higher the value, the better the insulation level. If you live in colder climates where the number of heating days is high, it may be in your best interests to increase these thermal resistance values.

As the price of heating fuel continues to rise over the coming years, any effort you make now to create a more energy-efficient home will be paid back several times over. I recall my parents' first home, built at a time when fuel costs weren't considered at all. This home was constructed using 2 x 4" lumber (63 x 125 mm) and minimal levels of insulation, with no wind barrier and no vapor barrier. But with heating-fuel costs in the $0.25/gallon range, no one cared.

How times have changed. With oil prices now ten times this price, no one can afford to let their heating dollars escape because of poor home planning or construction. Although no one has a crystal ball, I suspect that energy prices will rise considerably faster in the coming years. Our parents had little or no concern about supply, environmental issues, security, or hostile foreign governments. We do not have this luxury.

Table 3-1: Minimum Recommended Insulation Values

Insulation Quality (over unheated space)	Walls	Basement Wall	Roof	Floor
R Value	23	13	40	30
RSI Value	4.1	2.2	7.1	5

Moisture Barriers

A vapor barrier consisting of 6-mil-thick (0.006") polyethylene plastic can be attached to the wall structure on the warm side of the insulation. The vapor barrier completely surrounds the inside of the house and must be well sealed at the joints and overlapped edges. During construction, care must be taken to ensure that this barrier extends without a break from one floor to the next.

The function of the vapor barrier is to seal the house in a plastic bag, controlling air intake and leakage. It also stops warm, moist air from penetrating the insulation and contacting cooler air, condensing, and causing mold and rot problems in the wall. Additionally, a vapor barrier ensures that the air inside the insulation remains still. These conditions are absolutely essential in making the insulation system function properly.

Uncontrolled moisture can cause wood rot, peeled paint, damaged plaster, and ruined carpets. Moisture can also directly influence the formation of molds, which are allergens for many people. Moisture control is not to be taken lightly. Left uncontrolled, it can cause damage to building components.

Some insulation materials such as urethane foam spray or sheet styrofoam are fabricated with millions of trapped air or nitrogen bubbles in the plastic material. Although more expensive than traditional fiberglass, inch-for-inch they provide higher insulation ratings and also form their own vapor barrier.

Wind Barriers

A properly taped and well-jointed Tyvek®-style wind and water barrier on the outside of the house, just under the siding material, helps to keep winter winds from whistling right through the insulation. Controlling wind pressure infiltration into the building structure is an area most people overlook. Even if the breeze isn't blowing through your hair while the windows are closed, wind leakage is important.

Insulation works by keeping dry air very still. Even the slightest movement will greatly reduce the insulation value of the system. Think about how much force you exert wrestling with an umbrella in even a light wind. Now imagine the same effect wind has on the entire surface area of a house. Quite frankly, it's amazing more homes don't blow down! The force of the wind on the house structure can penetrate the insulation. This upsets our still air requirement and lowers insulation values.

Provided all of the above insulation techniques are followed carefully, you will have a very efficient home-insulation system.

Examples of these techniques are shown in the cross-section or side view of the bungalow in Figure 2-3. This design strengthens your local building code by increasing insulation in areas that you may not have considered.

Basement Floor

This floor comprises a 3" (75 mm)-thick sheet of extruded foam-board insulation installed directly on top of packed crushed stone. Over this can be applied an 8-mil layer of polyethylene moisture barrier, with all joints taped

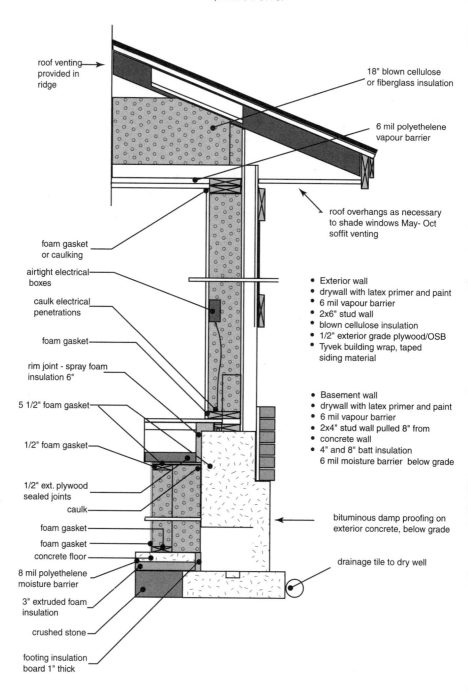

roof venting provided in ridge

18" blown cellulose or fiberglass insulation

6 mil polyethelene vapour barrier

roof overhangs as necessary to shade windows May- Oct soffit venting

foam gasket or caulking

airtight electrical boxes

caulk electrical penetrations

foam gasket

rim joint - spray foam insulation 6"

5 1/2" foam gasket

1/2" foam gasket

1/2" ext. plywood sealed joints

caulk

foam gasket

foam gasket

concrete floor

8 mil polyethelene moisture barrier

3" extruded foam insulation

crushed stone

footing insulation board 1" thick

- Exterior wall
- drywall with latex primer and paint
- 6 mil vapour barrier
- 2x6" stud wall
- blown cellulose insulation
- 1/2" exterior grade plywood/OSB
- Tyvek building wrap, taped siding material

- Basement wall
- drywall with latex primer and paint
- 6 mil vapour barrier
- 2x4" stud wall pulled 8" from concrete wall
- 4" and 8" batt insulation 6 mil moisture barrier below grade

bituminous damp proofing on exterior concrete, below grade

drainage tile to dry well

Figure 2-3. This cross-section of a high-efficiency house illustrates additional insulation and sealing techniques that will greatly reduce your heating and cooling energy usage.

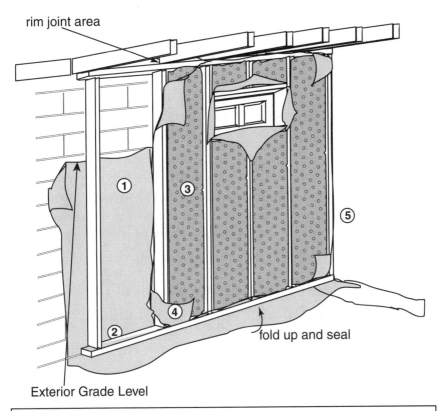

rim joint area

Exterior Grade Level

fold up and seal

Interior insulation involves: 1) a moisture barrier; 2) new frame wall; 3) insulation; 4) air and vapor barrier; 5) finishing

Figure 2-4. The technique illustrated provides excellent insulation value and ensures that your basement will also remain dry and mold free.

or overlapping seams caulked with acoustical sealant. Such a membrane prevents ground-based moisture from entering through the floor area and thus reduces mold potential. It will also stop ground gases including radon from entering. In areas where foundation footings are not subject to frost heaving, a footing insulation board can be added between the basement wall and the concrete slab floor.

Basement Walls

The basement walls can be "finished" using standard framing techniques, with the exception that the 2 x 4" stud wall is set away from the concrete basement wall by 4" (100 mm). This increases the wall cavity space for extra insulation without requiring additional framing material. Instead of traditional batt insulation, blown-in fiberglass or cellulose is suggested. These materials ensure complete coverage and packing density, especially in the hard-to-insulate areas around plumbing lines, electrical boxes, and wiring. Figure 2-4 shows the use of a tarpaper-style moisture barrier glued to the concrete wall from just below grade. It is then folded up and glued to the bottom of the interior vapor barrier.

Rim Joints

Rim joints are not an obscure arthritic condition, but rather an obscure area of your home where the floor joists meet with the exterior support walls. (They are also known as rim joists or header joints.) They are known to be notoriously difficult to insulate and vapor barrier correctly, given the huge amount of surface area accorded this space. Many insulation contractors give only passing effort to this area, stuffing a piece of insulation in place and then stapling a swath of plastic on top, as shown in Figure 2-5a.

A much better method of insulating the rim joists is to use a spray foam material. Although more expensive than traditional methods, this process actually works. As an added benefit, the foam also provides its own vapor barrier, as shown in Figure 2-5b.

The photograph on the left (Figure 2-5a) details a too typical "stuff and run" rim joist insulation job. The photograph on the right (Figure 2-5b) shows the best way to complete rim joist insulation using sprayed urethane foam, which provides an integral vapor barrier.

Exterior Walls

Typical framed construction now employs 2 x 6 " (50 x 150 mm) stud walls. This wall-cavity thickness is suitable in all construction areas provided quality workmanship is assured. The major problem with exterior wall construction is air leakage. (Have you ever felt the breeze blowing from the electrical boxes of older homes?) The use of blown-in fiberglass, cellulose, or rock

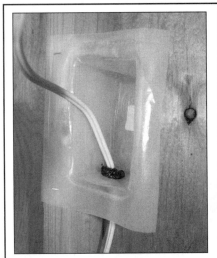 

Figure 2-6 Electrical boxes and other wall penetrations must be carefully sealed with the vapor barrier. You can purchase electrical boxes that are integrally sealed or fabricate your own using a piece of 6-mil polyethylene sheet.

Figure 2-7 The vapor barrier must create a "plastic bag" effect for the entire home. Ensure that this membrane is wrapped from one floor to the next and caulked at each joint.

wool in the wall cavity will ensure complete coverage over the entire surface area. This technique also ensures that insulation works its way around electrical boxes and other obstructions in the cavity. Make sure the contractor fills the wall with a sufficient density of insulation. The material should be tightly packed and not sagging at the top of the frame wall.

The next areas to consider are electrical, phone, and cable boxes, which penetrate the exterior structure. Ensure that airtight boxes are used and caulk between them with a 6-mil layer of polyethylene vapor barrier. The vapor barrier must be overlapped and caulked tightly at all seams and with the top and bottom plates making up the wall section.

Air can easily blow between the top and bottom plates of the wall and the floor components. For this reason, these areas should have foam gaskets or caulking compound to ensure air tightness.

Windows

It would be quite easy to write an entire book on the design and treatment of windows. For our purposes, we will consider windows just from an energy standpoint. A single pane of glass is a very poor insulator. During the last thirty years, there have been many advances in window design including double- and triple-pane versions, low-emissivity coatings, argon and krypton gas fillings, and a bewildering array of styles and construction.

At the risk of oversimplifying this issue, follow these basic rules when choosing your windows:

- Purchase the best-quality window you can afford. Ensure that it is from a reputable supplier and has a good warranty. Ask for references.

- If you don't require solar gain, purchase at least double-pane windows with low-emissivity glass (also known as "low-e"). This type of glass reduces heat gain and loss. Note, though, that low-e glass and coatings should not be used with south-facing windows where winter solar gain is desired.

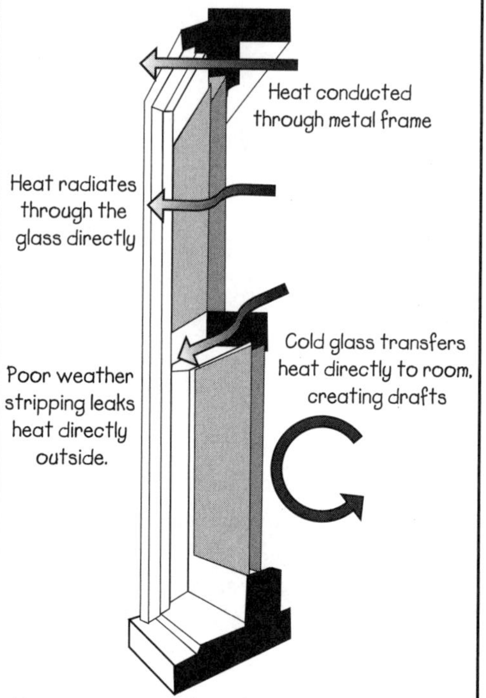

Heat conducted through metal frame

Heat radiates through the glass directly

Poor weather stripping leaks heat directly outside.

Cold glass transfers heat directly to room, creating drafts

Figure 2-8 Windows lose heat through air leakage as well as direct conduction through the glass and frame. Always purchase the best-quality window you can afford from a reputable company.

- Ensure that the cavity between window panes uses argon gas, or better still krypton gas. Triple-glazed windows using low-e glass and filled with krypton gas have an insulation value 3.5 times better than that of traditional double-pane windows.

- All framing materials used in the construction of the window should have wood or vinyl cladding to prevent heat transfer through a metallic structure.

- When installing windows be sure to adequately insulate and caulk, preferably with spray foam, between the window and the house frame.

- Make certain that windows are shaded with roof overhangs, awnings, or leaves of deciduous trees from early May to mid-October to prevent undesired solar heating. In turn, ensure that windows are free to capture the winter sun for the remainder of the year.

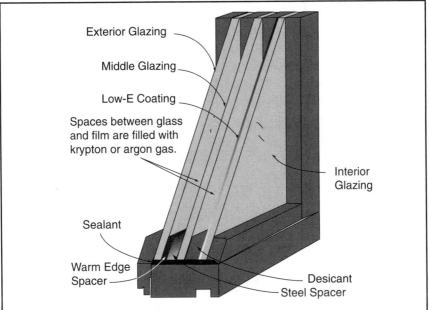

Exterior Glazing

Middle Glazing

Low-E Coating

Spaces between glass
and film are filled with
krypton or argon gas.

Interior
Glazing

Sealant

Warm Edge
Spacer

Desicant

Steel Spacer

*Figure 2-9 Modern high-efficiency windows use two or more panes of glass which
are separated by thermal insulation and filled with argon or krypton gas which
increases the insulation value. When solar gain is not a factor (such as on the north
side of the home), purchase windows with low-emissivity coatings.*

*Figure 2-10 Solar blinds such as these sunshades block 97% of the daylight heat
energy and glare while allowing you to see outside (www.shade-o-matic.com). Quilted
blockout curtains or bamboo shading blinds will also work.*

Attic and Ceiling Treatment

Provided the attic is not finished and has a suitable vapor barrier, the best way to insulate this area is to blow in 18" of cellulose or fiberglass. To ensure that the area above the insulation remains as cool and dry as possible during the year, install adequate roof and soffit vents to provide proper air circulation: warm air rises and exits through the roof peak vents while cooler outside air is drawn in through the soffit vents. This upgrade will eliminate moisture damage and ice damming in winter.

If you live in an area where summer air-conditioning load is a higher concern than winter heating, you may wish to consider radiant insulation. This material is similar to aluminum foil stapled to the underside of your roof rafters. Working like a mirror, the film reflects heat back through the roof surface before it gets a chance to hit the attic insulation.

The ceiling vapor barrier should continue down the wall surface a few inches and overlap the wall vapor barrier. Where these overlap, ensure that adequate caulking is applied. The best vapor barrier job is wasted if the finishing trades rip, cut, or damage the sealed home. We are striving to have you live inside a plastic bag, so take extra care to seal and repair any cuts. Good workmanship will guarantee an energy-efficient home.

Lastly, gasket and seal the attic hatch door in the closed position. This reduces air leakage around its perimeter.

If your home has more complex design features, be sure to discuss these general guidelines with your architect during the planning stage.

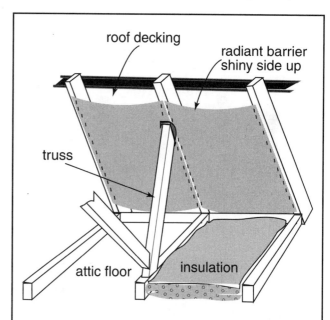

Figure 2-11 Summer air-conditioning loads can be reduced by the use of radiant insulation, which reflects heat back through the roof.

Ventilation Systems

Living in an energy-efficient house is like living in a plastic bag. With all of the air leaks sealed and the vapor barrier extending from the basement floor to the ceiling, we really have created a sealed, airtight home. Contrast this with the typical older house with leaky kitchen and bathroom fans and other cracks and leaks in the building structure. While these provide ventilation, they are completely uncontrolled and cause tremendous losses in heating and cooling energy. In a typical thirty-year-old house, it is estimated that these leaks collectively equate to a one-square-foot hole in the wall!

To ventilate your sealed home, you will have to resort to alternative measures. The methods you use will depend to a large extent on where you live.

Most high-efficiency homes utilize a device known as an air-to-air exchanger or heat recovery ventilator (HRV)—a marvel of simplicity which is brilliant in its execution. Stale, moist air from the kitchen, bathrooms, and other areas is drawn into the HRV unit and passes over a membrane before being exhausted from the house. At the same time, colder incoming fresh air is pulled into the unit and passes over the opposite side of the membrane.

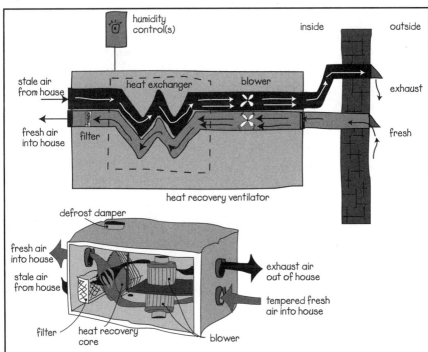

Figure 2-12 The heat recovery ventilator (HRV) controls air flow into and out of a house. Warm exhaust air is passed over a membrane, transferring this heat energy to the colder, incoming fresh air.

This causes waste heat to be transferred to the colder, dry incoming air, which is then warmed and distributed in the central furnace air duct.

Controls and timers located throughout the house can be programmed to monitor humidity and smoke. Even carbon dioxide levels can be monitored, indicating that the house is occupied and adjusting ventilation accordingly. Based on rules programmed into the control system, the HRV automatically adjusts air-exchange flow to current conditions, while at the same time saving plenty of heating dollars.

Homes in areas that have a low heating load in winter do not fully benefit from the heat-recovery aspect of the HRV. However, these homes would benefit from the ability of the HRV to filter dust and pollen as well as provide proper ventilation.

Heating Controls

Have you heard the old story about conserving energy by turning your thermostat down when going to bed at night and then turning it back up in the morning? It is a story because no one really does it, at least not for long anyway. Besides, who wants to wake up to a cold house?

Add a little technological marvel to the furnace control. A setback thermostat will take care of reducing the furnace or air conditioning output while you are sleeping or at work. When you wake up or return home, the house will be at the de-

Figure 2-13 The setback thermostat is simplicity itself. Program your daily schedule into the unit and the thermostat will automatically adjust the heating and air-conditioning system to your personal schedule.

sired temperature. Without even being aware of it, you'll save money and reduce greenhouse gas emissions. Most models even know the difference between weekdays and weekends, providing you with lots of customization capability.

New Construction Summary

By following the tips provided above, you can achieve a satisfactory balance between capital cost and energy conservation. In contrast to a home built to standard building codes, a home designed using the principles provided in this (and the next) section can reduce your heating and cooling energy requirements and costs by 50% or more.

2.2
Updating an Older Home

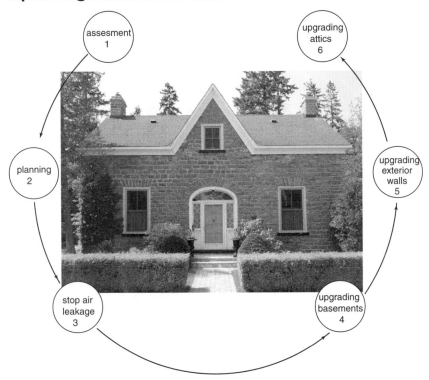

Figure 2-14 Older homes have incredible character and appeal. What they lack is energy efficiency. Proper assessments, planning, and retrofit work always make energy sense.

If your home is more than ten years old, it is likely that many of the energy-saving features described in the previous section are not incorporated into its design. Don't despair; there is no need to tear down the old place or move. Instead, we can review the entire home systematically and determine where to put your renovation dollars to make energy sense.

Where to Start?
Every home is unique and, whether it's one, ten or a hundred years old, what works for one house may not work for the next. As with any project, an assessment is the place to start. We need to check the general condition of the house and test what areas need to be beefed up. Over the last few decades, home designs have improved and architects, engineers, and building

contractors have all learned from past practices. With this in mind, there is a general checklist we can use to determine what energy shape our house is in right now.

Step 1 – Assessment

- Consider having a professional inspection of your home. Programs such as the Canadian *EnerGuide for Homes* assessment will not only identify problem areas but also provide suggested corrective action and even precalculate government rebates for implementing the suggested work.
- Air leakage is the most common problem with all older homes. Poorly fitting window frames and leaks around doors and chimneys abound. Simple corrective actions such as applying sealants and weather stripping stop breezes from blowing into the house. Correcting air leakage also helps to regulate humidity and reduce condensation problems in existing insulation.
- Many older homes treat the basement like an outcast relative: it's there, but leave it alone. Often dark, damp holes which are not properly insulated, basements can eat an enormous number of heating dollars. Moisture problems tend to be left to a dehumidifier, which is about as useful as putting a bucket under a leaky roof. Let's try to correct these issues with damp-proofing, moisture barriers, drain gutters, and proper insulation systems.

Figure 2-15 A professional assessment of your home will include a pressure door test. A computer-controlled fan will pressurize a home, calculate air leakage, and allow the assessor to identify problem areas in the home and suggest corrective action.

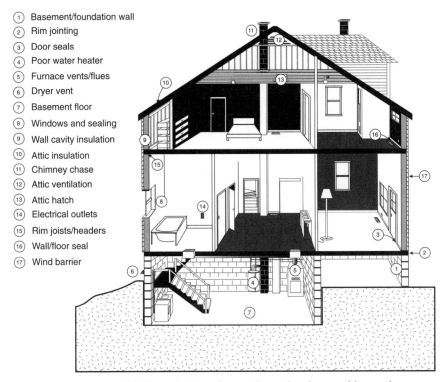

1. Basement/foundation wall
2. Rim jointing
3. Door seals
4. Poor water heater
5. Furnace vents/flues
6. Dryer vent
7. Basement floor
8. Windows and sealing
9. Wall cavity insulation
10. Attic insulation
11. Chimney chase
12. Attic ventilation
13. Attic hatch
14. Electrical outlets
15. Rim joists/headers
16. Wall/floor seal
17. Wind barrier

Figure 2-16 A thousand little air leaks and a poorly insulated area add up to big energy costs. A thorough assessment of these areas will let you know where you are wasting your money.

- Cavities between walls in older homes are woefully underinsulated or completely uninsulated. Stone walls look great on the outside, but are murder on the heating bill if not properly dressed inside. There are many ways of tackling these problems, from both the inside as well as the outside of the home.

- Everyone knows that heated air rises. Let's try to stop it before it escapes through underinsulated attics, leaky joints, and chimney and plumbing areas.

Air Leakage

Air leakage control is the most important step that can be taken in upgrading any home. Controlling air leakage provides many benefits:

- Heating/cooling costs are reduced as the infiltration of outside air is decreased.

- Insulation efficiency is increased, since still air allows insulation to work properly.

- Humidity and condensation levels can be controlled.
- Home comfort is improved as drafts and cold spots are eliminated.

It is possible to hire professional contractors to assess the quality of the air barrier system in your house. Alternatively, a simple yet effective method is to conduct the test yourself using several incense sticks left over from your psychedelic days in college.

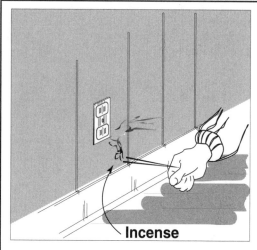

Figure 2-17 A burning incense stick held near suspected areas of leakage on a windy day will cause the smoke to move towards or away from the leak.

Hold the incense near the suspected leakage area on a windy day and observe what happens to the smoke. Smoke drawn towards or blown away from a suspect area indicates an air leakage path. Mark the location down on your list. Continue testing in this manner, paying attention to:

- electrical outlets, including switches and light fixtures
- plumbing penetrations that include the attic vent stack, plumbing lines to taps and dishwashers, etc.
- floor-to-ceiling joints and other building areas that are storied
- baseboards, crown

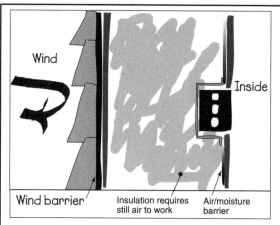

Figure 2-18 Proper wind and air barriers combined with adequate insulation keep the heat in. If any one element is missing or faulty, you may be throwing money away.

molding, and doorway molding
- fireplace damper area and chimney exit through the attic or wall
- attic hatch
- windows and doors (Ensure that glass fits tightly and casing area is sealed.)
- ventilation for appliances (kitchen and bathroom fans, dryer vents, gas stoves, water heaters, etc.)
- pipes, vents, wiring, and plumbing lines in the basement and attic (Move the insulation if necessary to access these areas.)

In an older home, it is quite likely that the smoke from the incense will blow just standing in the middle of the room. Don't despair. Just mark all of the areas down on your sheet and perhaps record a "severity rating" from 1 to 10. This way you can tackle the tough problem areas first.

The Basement

Most homeowners don't even consider unfinished basements as a source of heat loss. Part of this mentality comes from the mistaken idea that heat *only* rises and that earth is a good insulator. Both are wrong. Heat travels in any direction it chooses, but it always travels from a warm area to a colder one. Additionally, older basements have large surface areas of uninsulated walls and flooring that act as a heat sink, drawing warm air to these cooler surfaces.

As we discussed in the *Air Leakage* section above, there is also a lot of heat loss through crevices in the walls, around and through windows, and at the top of the foundation wall where it meets the first floor. An uninsulated basement can account for up to 35% of the total heat loss in a home.

It is not possible to simply add insulation and air barriers to a damp, leaky basement without first correcting any underlying problems. Any areas that accumulate water in the spring or after a heavy rain must be repaired, as wet insulation has no energy-efficiency value and will ultimately contribute to mold and air quality issues. Check the basement for dampness, water leaks, and puddles in wet periods, and look for major or moving cracks in the foundation wall. Also ensure that a sump pit and pump have been installed in areas where persistent water accumulation occurs.

Some basement wetness problems are corrected by sloping the landscaping away from the foundation wall or adding rain gutters to the house eaves. If you are not sure that these measures will correct the basement water problems, it will be necessary to excavate around the perimeter of the house to provide supplementary drainage and to apply damp-proofing and external, waterproof insulation treatments to the foundation wall.

Wood-Framed Walls

Wood-framed walls are the easiest to insulate, as they are very similar to new construction. The major concerns relate to wall-cavity thickness and access. When assessing these wall structures, attempt to determine if the wall is empty or if there is already some form of insulation in place.

The simplest means of checking these walls for insulation is to remove the cover plate of an exterior-wall electrical plug or switch and, using a flashlight and a thin probe, check for insulation behind the electrical box. CAUTION! Before completing this test, turn off the circuit breaker or remove the fuse for that circuit and test the outlet or switch to verify that it is off. This test will have to be conducted at a few points around the house to ensure that your sample investigation is accurate.

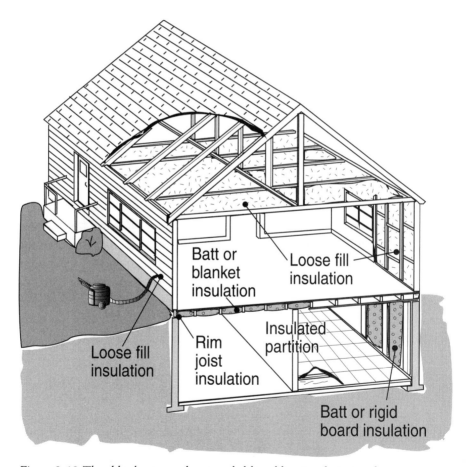

Figure 2-19 The older home can be upgraded by adding insulation to the structure as shown above. Air and vapor barriers may also be added.

Brick Walls

Brick walls of homes are almost always constructed of veneer with a frame wall on the interior side. Usually there is a small air gap between the brick and the frame wall to allow air to circulate and prevent moisture on the brick from rotting the framing members.

This air gap must not be filled with insulation. However, the frame wall might allow for insulation to be added. Use the same tests as those used for a standard frame wall to determine the depth and area that may be insulated.

Stone or Other Solid Walls

Stone, concrete block, cut stone, and other solid-wall treatments are similar in nature to brick walls. Solid wall treatments are not suitable for insulating and sealing and will require extensive reframing on the inside to add an insulation cavity and vapor barrier. Alternatively, special insulating siding and coverings may be added to the exterior.

Attics and Roof Areas

Most homeowners love to dump their home insulation upgrade dollars in the attic. Perhaps this results from the mistaken idea that all heat rises and the losses must therefore be highest in the attic? Could it also be because attics are one of the easiest places to insulate first? Just dump a few bags of insulation in the old attic and, *voilá!*, your heating bill goes down by 50%?

Sorry, not so fast. The attic does lose heat, but it actually loses less than an uninsulated basement or exterior wall. Most homeowners, no matter how old their house, may have had a passing thought about attic insulation and may have even added a few inches of something up there. Adding more insulation on top of old is not a problem unless the air leakage tests discussed earlier confirm that problems exist. If this is the case, it may be necessary to remove or move aside existing insulation to gain access to the leaking area. The necessity of ensuring quality air sealing in the attic area cannot be understated.

Another potential problem in the attic is moisture, which may originate from different sources such as a leaky roof, ice damming, or frost. Moisture in the attic can also come from within the house through leaky ventilation fan outlets from the bathroom and kitchen areas.

Ventilation of the attic itself is important to provide summer cooling and winter dryness, but many older homes do not have adequate (or any) vents. Ventilation is provided by air intake vents in the soffit and roof vents in the gable or peak where hot air can exit. There should be a ratio of 1 unit of roof vent area for every 300 units of attic floor area. For example, a 1200-

square-foot attic should have 4 square feet of vent area.

A common upgrading practice is to add electric vent fans to the attic area, usually in the gable or roof peak. This is not required, nor is it recommended. An electric exhaust fan increases airflow in the attic and may exceed the intake capacity of the soffit vents. If this should occur, additional air will be drawn from the main part of the house, which is exactly the opposite of what is desired.

Check the attic several times during the year, for example after a heavy rain or on a very cold day. Look for wet areas, mold, and rot or small "drip holes" on the insulation or attic floor surfaces.

Attics come in all manner of shapes, sizes and designs. The trickier it is getting access to the existing insulation, (or locating where new insulation should be placed) the more difficult it will be to upgrade, and therefore professional assistance may be required.

Step 2 – Planning the Work

You may wish to tackle some of the upgrading work yourself. Most of the tools required are pretty common household items and the few specialized tools can be rented from your local rental depot or borrowed from a friend. One couple recently completed a blown-in cellulose cathedral ceiling while balancing on a scaffold and using an insulation blower loaned to them by the material supplier. The do-it-yourself approach might not be for everyone, but if you do it correctly, it can result in considerable savings.

Some upgrading work is best left to the professionals: urethane foam spray insulation, for example, requires a truckload of specialized equipment; and if excavating around the foundation of the house is a job that just doesn't make the top of your "list of things to do in life," a backhoe and an experienced operator will really work wonders.

Building Codes

The various national, state or provincial, and local building codes as well as their variations are enough to frustrate any do-it-yourselfer. But don't cut corners. The reason the codes are there in the first place is for your health and well-being. Get to know your local building official. Although there are plenty of horror stories circulating about these officials, they are generally started by people who began the upgrade work before completing the planning and getting a building permit. You will find that most officials are very helpful, providing guidance on technical issues and referrals to qualified contractors or suppliers in your area. Work *with* them.

Safety

Working around the house climbing ladders or working with insulation and chemicals can be dangerous. Make sure you have the proper safety equipment including work boots, dust masks, rubber or latex gloves, and eye protection. Attic and basement areas may not be well illuminated; ensure that you have a suitable light. Use caution on the attic "floor." Often this floor is nothing more than the drywall or lath and plaster finish material on the ceiling below. It will barely hold a cat, let alone your body weight.

Step 3 – Stopping Air Leakage

Now that we are armed with our "list of air leaks" from the assessment phase, we can start getting down to business:

Caulking

Seal up small cracks, leaks, and penetrations on the inside (warm side) of exterior walls, ceilings, and floors. Sealant applied on the inside lasts longer as the material is not exposed to the elements outside.

Caulking is done using an inexpensive gun with a tube of appropriate material. There are literally hundreds of caulking materials available. Discuss with a building supply store the type best suited to your project. After a caulking job, many people are dissatisfied with the brand they used and/or the job they did. Avoid the tendency to purchase poor-quality materials which are difficult to apply and do not last.

Remember to purchase high-temperature silicone or polysulfide compounds for areas around wood-stove chimneys or hot water heater flue vents.

- Identify the area to be caulked. Determine the compound type appropriate for the job.
- Never caulk in cold weather. Caulking should be applied as close to room temperature as possible.
- Clean the area to be caulked. Large cracks and holes greater than 1/4" will require a filler of oakum or foam rope sold for this purpose.
- The nozzle of the compound should be cut just large enough to overlap the crack. Insert a piece of coat hanger wire or a long nail into the nozzle to break the thin metal seal.
- Pull the caulking gun along at right angles to the crack, ensuring that sufficient compound is dispensed to cover both sides of the crack. Remember that caulking shrinks, so it's better to go a bit overboard.
- The finished caulking "bead" should be smooth and clean. The surface of the bead may be smoothed with a finger dipped in water.

- Some compounds require paint thinner or other chemicals to clean up. Check the label of the compound before using it to determine the cleanup method.

Electrical Boxes

Air leakage around electrical boxes is so common that there is an off-the-shelf solution available. Special fireproof foam gaskets and pads may be added just behind the decorative plate as shown

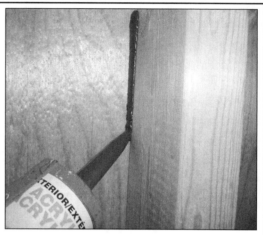

Figure 2-20 It takes a bit of practice to make a good caulking joint. Go slowly and don't cut the nozzle too large; otherwise you'll end up with hard-to-clean goop!

in Figure 2-21. To ensure a superior seal, use an indoor latex caulking compound on the gasket face before applying the decorative cover.

If the room is being renovated at the same time, install plastic "hats" around the entire electrical box. These hats have an opening for the supply wires and small flaps that can be sealed to the vapor barrier. Ensure that the area where the supply wire enters the hat is well sealed with acoustic sealant.

Windows

Older homes often have single panes of glass puttied into wooden frames. When the putty dries out, air leakage occurs. Remove old putty and replace it with glazing compound to ensure a high degree of flexibility and a long life.

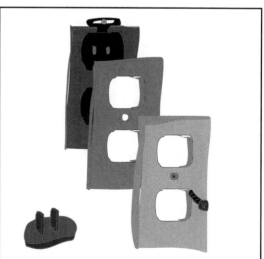

Figure 2-21 Air leakage is so common in older homes that off-the-shelf sealing pads have been designed to aid in stopping leaks.

Old-fashioned putty is not recommended.

The area between the window and the frame is another area prone to leakage. Access to this area usually requires the removal of the window casing trim. Use oakum, foam rope or, better still, urethane foam spray to air/vapor seal and insulate this area in one application.

Baseboards, Moldings, and Doorway Trim

Trim pieces such as baseboards, moldings, and door casing are used to cover the gap between one framing section and another. For example, a premanufactured door is placed in a framed section of the wall called the "rough opening." Obviously the door must be smaller than the opening in order to fit into it. If this opening is not properly sealed, considerable air leakage will result.

The best way to seal these areas is to use methods similar to those described for window framing (see Figure 2-22). For smaller gaps or areas where urethane foam may be too messy (near carpets or finished wood), oakum or foam rope may be jammed into the gap.

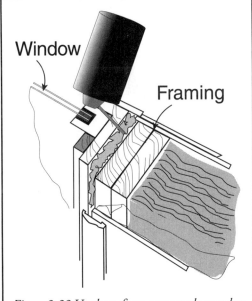

Fireplaces

Fireplaces warm the heart, but rob you blind. Air leakage up and down the chimney when the unit is not in use is enormous. When a fire is burning, the suction it creates draws large volumes of expensive, heated room air

Figure 2-22 Urethane foam spray works wonders in small nooks and crannies such as those between the window and house framing.

up the chimney, drawing in cold outside air to replace the heated air that literally went up in smoke!

What to do? Simple. Replace the fireplace with an airtight wood-burning stove or a similar controlled combustion unit. We will discuss these items in greater detail in Chapter 3.

If you really must keep the fireplace, make a removable flue plug that properly seals the chimney when it is not in use. Check to make sure the damper closes as it should. If you detected air leakage around the chimney and framing as part of your assessment in Step 1, seal any cracks with heat-resistant sealant and mineral wool or fiberglass batting.

A quick word on glass fireplace doors; don't waste your money. If you like the look of them or are worried about flying sparks, so be it, but be warned that the majority of these door units are cheaply made and will not provide any air sealing capacity.

Attic Hatch

Seal the attic hatch in the same manner as any exterior doorway trim. Use hook and eye screws to ensure that the hatch fits snugly against the weather stripping (see Figure 2-23).

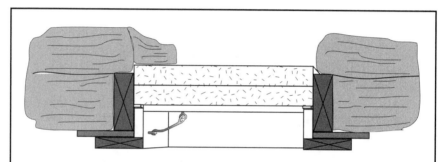

Figure 2-23 The attic hatch is often forgotten when it comes to air leakage and insulation. Use exterior doorway sealing gaskets and hook and eye screws to ensure that the hatch door fits snugly, preventing air leakage.

Air and Vapor Barrier

So far in our discussion of air leakage we have been concerned with the worst but most easily corrected areas. If you are undertaking more extensive renovations, it may be possible to improve large sections of the air and vapor barrier. Before we continue with how to perform this type of major upgrade, let's quickly review the function of each component to be sure we understand how this work can be implemented (see Figure 2-18).

On the exterior of the house, directly under the siding material, is a layer of spun-bonded olefin, which is often sold under the trade name Tyvek®. This material stops wind from penetrating into the insulation area and repels rainwater. Interestingly, it is also permeable to water vapor, which allows trapped moisture in the insulation cavity to escape outdoors, preventing rot.

On the warm side of the insulation is the vapor barrier. In newer homes, this barrier consists of 0.006" (6-mil) polyethylene sheet (think food wrap on steroids) affixed to the wall studs and carefully sealed.

The vapor barrier acts as a secondary wind barrier, stilling the air trapped in the insulation. It also prevents warm, moist household air from penetrating into the insulation. (Remember—heat moves toward colder areas.) Moist air contacting cold insulation and wood siding on the exterior side of the insulation will quickly condense, forming water. Should this water be present for extended periods of time, it will cause structural rot and mold problems. Many a house has been seriously damaged due to high inside humidity coupled with an inadequate vapor barrier.

Your Options

During renovation, it may be possible to install both an air barrier and a vapor barrier. This type of upgrade can only be completed when the framing members are accessible, which typically occurs when a major rebuilding effort is in progress or an addition is being constructed. Major renovation involving more than just a few sheets of vapor barrier is discussed in *Step 5 – Upgrading Walls*. It doesn't matter if your renovation work is concentrated on the exterior or interior side of the wall; in either case there are ways to complete the work.

Exterior Wall Upgrade

If you are upgrading, removing, or adding to existing siding, it is possible to increase the air barrier with little difficulty. Removal of the old siding reveals the wood sheeting covering the framing members. A layer of Tyvek® air barrier can be taped to the exposed wall sheeting or over existing smooth

Figure 2-24 A home wrapped with a suitable air barrier decreases air leakage and increases insulation values.

siding materials. The air barrier is similar to the vapor barrier in that all joints should be carefully taped to ensure a continuous barrier to wind penetration.

If the interior wall is not being upgraded, follow *Step 3 – Stopping Air Leakage*. Although a polyethylene film is not being applied, a reasonably good vapor barrier can be made using multiple layers of latex paint over well-sealed drywall. When an external air barrier is mixed with air-leakage sealing and latex paint, you can achieve the next-best thing to new construction.

Interior Wall Upgrade

Your choices for upgrading air and vapor barriers increase if you are updating an interior wall. We will discuss this work further in *Step 5 – Upgrading Walls*.

Step 4 – Upgrading Basements

In the section on basements in *Step 1*, we discussed the need to ensure complete dryness before insulation and/or vapor barriers can be added to the inside. Persistent moisture or water leakage problems will ruin even the best-quality work.

Dampness

Minor dampness causes staining or mold growth, blistering and peeling of paint, moldy smells, and efflorescence (whitish deposits) on concrete. These problems can be corrected from the inside by cleaning up any mold and then applying damp-proofing to the foundation wall. More serious problems will have to be corrected from the outside.

Figure 2-25 Installing rain gutters and sloping the grade away from the house are two simple but effective means of keeping the basement dry.

Major Cracks

If your foundation wall has large cracks or cracks that are getting bigger, seek professional help prior to upgrading to determine if structural repairs are necessary.

What Are My Choices?

There is no doubt that insulating from the inside of the basement is the easiest and least costly method, provided that the basement is dry. Insulating from the outside will do the best job from a technical point of view. What to do? Let's begin by weighing the pros and cons of each method.

Insulating Inside

Insulating inside usually involves the addition of a wood-framed wall and adding some form of insulation material, just like a standard wall. There are several advantages of insulating inside:

- The work can be done at any time of the year.
- A completed job will increase the value of your home.
- Indoor finished space will be increased.
- It is the lowest cost way to update the basement insulation and vapor barrier.
- The landscaping, porches, walkways, and other obstructions outside will not be disturbed.

There are also some disadvantages:

- Persistently damp or wet basements cannot be upgraded.
- Furnaces, electrical panels, vent pipes, and plumbing obstructions make do-it-yourself framing more difficult.

Insulating Outside

Insulating outside the home involves more complex and expensive excavation but at the same time provides an opportunity to correct existing foundation defects. There are several advantages of insulating outside:

- The outside wall is generally straighter and simpler to insulate once the excavation has been completed.
- Most moisture and water leakage problems can be corrected at the same time.
- Foundation cracking and other damage can be inspected and repaired.
- There is no lost space inside the home as a result of the added wall thickness.
- The weight or mass of the foundation is on the warm side. This mass absorbs heating and cooling energy, helping to balance temperature fluctuations.

There are also some disadvantages of insulating outside the home:

- Excavation work is costly and may be difficult if there are porches, finished walkways, or decks abutting the foundation wall.

Insulation	Type	R-Value (approx.)	R.S.I Value
Batts	**Fiberglass** or **Rock Wool**	3 1/2" (R-11) 5 1/2" (R-19) 9 1/2" (R-30)	90mm 1.9 140mm 3.3 240mm 5.0
Loose Fill	**Fiberglass** or **Rock Wool**	R-2.7 per inch	1.9 per 100mm
	Cellulose	R-3.7 per inch	2.6 per 100mm
Rigid Board	**Expanded Polystrene (Beadboard)**	R-4 per inch	2.8 per 100mm
	Extruded Polystrene	R-5 per inch	3.5 per 100mm
	Polyurethane or **Polyisocyanurate**	R-7 to R-8 per inch	4.4 - 5.6 per 100mm
	Polyurethane	R-7 to R-8 per inch	4.9 - 5.6 per 100mm

Table 2-2 Common Insulation Materials and Their Heat-Resistance Values in R/RSI.

- Storing the excavated dirt will damage lawns and bring mud and sand into the house.
- Work cannot be done economically in the winter season.

Once you have determined which system to use, the next step is to examine how the work should be done.

How to Insulate Outside
Step 4 - 1 - Preparation
- Prior to beginning the insulation work, remove any outside features that will get in the way. This includes decks, stairs, trees, walkways, and so forth.

- Determine where power, water, sewer, septic tank lines, gas, telephone, and other services enter the building. Contact your utility to locate unknown pipes and wires; this is usually a free service.
- Determine where the excavated dirt will go. Placing a polyethylene sheet or tarp on the grass to hold the dirt will make the cleanup easier.

Step 4 - 2 – Excavation

- USE CAUTION! The soil may be unstable and fall back into the excavation, causing injury or death.
- The excavation must extend down to the top of the footing
- Never dig below or near the base of the footing (see Figure 2-26, Item 8), as this will cause the foot to sink and the house to drop.
- USE CAUTION! Older rubble stone walls may collapse without the support of the surrounding soil. Seek expert help if you are in doubt.

Step 4 - 3 – Preparing the Foundation Wall

- Brush and fully clean the foundation wall. Scrape any loose concrete or rubblework from the wall (Item 2).
- If concrete is missing or damaged in places, it is best to apply a coating of parging (waterproof masonry cement) to the damaged or missing sections. Allow this to dry.
- Apply damp-proofing compound to the foundation wall (Item 7). Damp-proofing is a tar-like substance that is painted on the foundation wall from the top of the footing to grade level.
- Inspect the footing drain system (Item 4). If there is no footing drain, determine whether one can be added. The footing drain is made by installing a flexible pipe that has been manufactured with thousands of holes along its length. The pipe is normally "socked," with a pantyhose-like liner to prevent sand and dirt from entering.
- This pipe is laid down adjacent to the footing without affecting the undisturbed soil in this area.
- A 12" layer of "clear stone" is spread on top of the footing drain.
- A strip of filter fabric or "gardeners' cloth" should be applied on top of the clear stone to prevent sand and dirt from plugging the footing drain.
- The footing drain should be routed downhill away from the house to a drainage ditch or dry well. A dry well can made simply by excavating a hole 3-4' in diameter to a depth lower than the house footings. Place the end of the footing drainpipe in this hole and fill with crushed "clear stone" to within 12" of the top. Cover the top of the dry well and supply ditch with soil to finish.

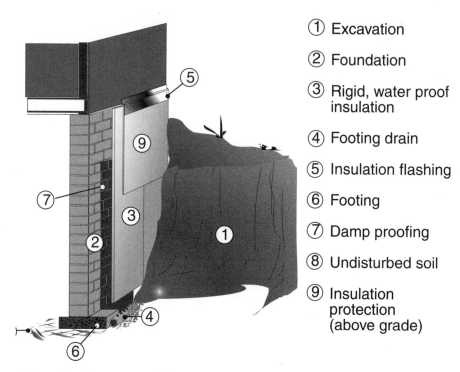

① Excavation

② Foundation

③ Rigid, water proof insulation

④ Footing drain

⑤ Insulation flashing

⑥ Footing

⑦ Damp proofing

⑧ Undisturbed soil

⑨ Insulation protection (above grade)

Figure 2-26 Excavation of the outside foundation wall is the best way to solve water leakage and foundation structural problems while upgrading insulation.

- Check that all services penetrating the foundation wall are well sealed with a suitable caulking compound. Remember to allow room for the insulation.

Step 4 - 4 – Installing the Insulation
- There are many types of insulation in use today, but the most common and easiest to work with is polystyrene rigid board. This material is supplied in sheets that are typically 2' x 8' long and have interlocking grooves running the full length. As indicated in Table 2-1, the minimum recommended insulation level of rigid board is R13/RSI 2.0. Your building supply dealer will be able to recommend which brand of insulation is available in your area. Just make sure you explain your R/RSI rating requirements.
- Rigid board insulation is very fragile and will break when exposed to wind or undue flexing. Make sure the board is protected prior to and during the installation phase of the work. A special flashing trim is in-

stalled at the top of the foundation wall that clips the insulation board in place. Alternatively, pressure-treated plywood may be applied on-site to hold the insulation to the header joist at the top of the foundation wall (Item 5).

- Make sure that all insulation joints are well sealed and clipped together. It is important that you discuss your specific needs for a flashing and insulation-clip system with your material supplier as there are many variations available.
- Ensure that insulation overlaps at the wall corners, preventing areas of concrete from being exposed.
- A covering is required to protect the insulation from sunlight and damage from traffic and animals where it protrudes above grade (Item 9). Coverings may be purchased for the application or fabricated from any of the following:
- pressure-treated plywood
- vinyl or aluminum siding to match the house
- metal lath and cement parging

Step 4 - 5 – Backfilling the Excavation
- After backfilling the drainage pipe as discussed in Step 3, it is time to refill the excavation. If the soil that was removed earlier is heavy clay or drains poorly, it is better to remove it and use clear-running "pit run" sand as the backfill material. This will greatly assist in encouraging water to drain away from the foundation wall, further ensuring a dry basement.
- When the excavation is refilled, ensure that the finished grade slopes away from the house. This promotes drainage and allows runoff to move away from the foundation wall. This is a good time to remind you to install eavestroughs with downspout pipes that lead away from the house.

How to Insulate Inside the Basement
Insulating a basement that is known to be dry is not much different than insulating a new house. The major differences relate to the type of wall structure that is used. A fairly new poured-concrete or block wall should present few problems. Older rubble or cut stone walls tend to be "wavy" and vary in height somewhat. This makes framing more difficult, but does not change the methods involved. The dry-basement insulation method is just a repeat of new-home construction:
- A tarpaper-style moisture barrier is glued to the concrete wall starting

from just below grade and then folded up and glued to the bottom of the interior vapor barrier as detailed in Figure 2-4.

- The basement wall is built using standard framing techniques except that the 2 x 4" stud wall is "pulled" away from the foundation wall by 4" (100 mm). This increases the wall-cavity space for additional insulation without using additional framing material.

- Instead of traditional batt insulation, blown-in fiberglass or cellulose is suggested. These materials ensure complete coverage and packing density, especially in the hard-to-insulate areas around plumbing lines, electrical boxes, and wiring.

- All other finishing details are exactly the same as in a new home.

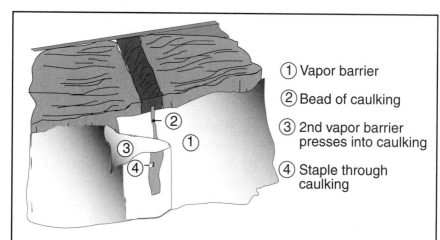

① Vapor barrier

② Bead of caulking

③ 2nd vapor barrier presses into caulking

④ Staple through caulking

Figure 2-27 A vapor barrier is installed in large continuous sheets. Where the sheets meet, they must overlap and be well sealed with acoustical sealant.

Problem Basements

There are a number of obstacles that can be encountered in older buildings, including:

- packed–stone or dirt floors;
- crawl spaces or basements with very low clearance heights;
- no-basement, slab-on-grade construction;
- building is on piers or blocks.

While it is possible to insulate problem basements, you should seek a professional contractor to review your specific requirements. There are too many variations in climate and construction type to generalize on how to deal with each scenario.

Step 5 - Upgrading Exterior Walls

Because of the large surface area they cover, walls account for a sizeable percentage of the heat loss in houses. Older homes, constructed of solid material such as stone, brick, or log, gave little thought to interior insulation. Many of these designs have a small air space between the exterior covering and the small inner frame wall. This area must not be insulated, as it is used as a drainage cavity for water leakage and condensation.

Hollow concrete-block walls should not be filled with insulation. The quality of the insulation and the "thermal bridging" effect of heat and cold passing through the block does not warrant the trouble or expense.

Traditional frame walls are easily insulated as there is usually an accessible cavity. Using various construction techniques, you can determine if there are cables, duct work, or other obstructions inside the wall cavity that may interfere with the application of insulation.

Solid walls of stone, brick, or log may be insulated from either the inside or outside, depending on several factors:

- The building may have heritage appeal. Homes built of traditional stone, log, or brick may be too beautiful to cover. Insulating from the inside may make the most sense in these situations.
- The outside may need refinishing. If your exterior siding or building material is looking a little tired, there are a number of ways to insulate from the outside and refinish the exterior siding at the same time.
- Interior walls may need new lath and plaster or general updating. When the inside wall surface is cracking and the wallpaper is starting to get to you, perhaps insulating from the inside is the right choice.
- Some exterior and interior finishing may be required. If energy efficiency is the ultimate quest, why not consider doing both? It is possible to add insulation to both the outside and inside

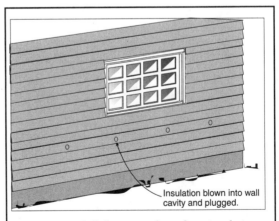

Insulation blown into wall cavity and plugged.

Figure 2-28 Cellulose or urethane foam insulation may be blown into the wall cavity from either the interior or the exterior side of the wall.

exterior walls, creating a "good as new" insulated home.

Construction type, material costs, and your skill level will determine the path you take on this upgrade.

Frame-Wall Cavities

Frame-wall cavities are by far the easiest to upgrade provided they are empty. A wall that is half-filled with insulation from an earlier job makes it almost impossible to do a good job and ultimately will not be worth the expense.

During the assessment stage, we determined how much insulation was in the wall by using a flashlight and poking around through electrical outlets (with the power off!) and other access points. If there is little or no existing insulation, a contractor will apply either cellulose fiber or polyurethane foam into the cavity. These materials are applied through small holes drilled into either the exterior siding or the interior drywall finish.

The holes drilled into exterior siding are plugged using wood dowels. Once the siding is repainted, the work is almost invisible. Brick homes that are suitable for blown insulation have selected bricks removed. The insulation is sprayed and the bricks are replaced.

If the work is being done from the inside, a hole is drilled into the drywall or lath and plaster surface at strategic locations. After the insulation is applied, the holes are filled, primed, and repainted.

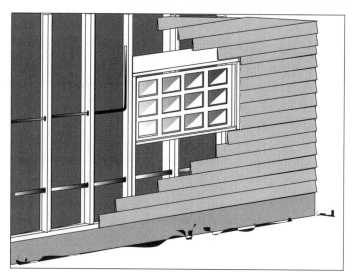

Figure 2-29 Make sure your contractor checks for obstructions or blockages in the wall, including window and door frames.

If the work is being done inside and the interior drywall or lathwork is poor, consider drilling holes into the existing surface and then covering it with a full vapor and air barrier. New drywall can then be placed directly on top of the existing surface. This technique uses up very little space and is cheaper than a fully framed interior wall.

Upgrading Insulation on the Exterior

If it's time to upgrade the old siding, this is an excellent opportunity to upgrade exterior insulation as well. There are several methods available that allow you to add insulation under the new siding, significantly increasing the overall "R/RSI" value of the home. If you are applying new siding, using blown-in insulation will eliminate the need for filling and repainting access holes. Consider these points as well:

- It is possible to add a generous amount of insulation by using high-density rigid insulation sheets or creating a new wall cavity on top of the existing siding.
- If the house is poorly insulated or made of stone, solid brick, or masonry, add a vapor barrier directly to the inside of this surface, under the new insulation. Follow the rules outlined above regarding proper vapor barrier installation and sealing.
- Remember that doors and window openings will have to be extended to allow for the additional insulation thickness.
- Ensure that water runoff from the roof will not drip between the old and new wall, ruining the insulation and causing structural rot damage. An eaves extension or flashing will prevent this.
- Make sure that the new insulation is well air-sealed. There is not much sense in doing a great insulation job with the winter winds howling between the old and new work.

Applying Exterior Insulation

There are dozens of ways exterior insulation can be applied. We will review the most popular methods and then point you to your local building supply contractor for more detailed information about materials for your specific application. Regardless of the application method you choose, be sure to wrap the entire upgrade work in spun-bonded olefin (Tyvek® brand) air barrier and ensure that it is carefully wrapped and taped at all joints.

1. It is possible to add exterior insulation by simply purchasing insulated siding materials. This is the easiest approach, although the insulation values are somewhat limited owing to the small thickness. Check with your building supplier on the many finishes available. These preinsulated

siding products install in the same manner as their uninsulated counter-parts.

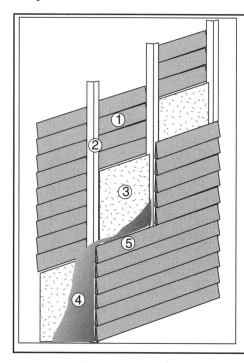

① Existing siding

② Framing members

③ Rigid or blown-in insulation

④ Air barrier

⑤ New siding

Figure 2-30 There are many ways to add exterior insulation to your home. Wood strapping used to support either rigid or batt insulation works very well.

2. Rigid board insulation can be applied directly over existing surfaces using appropriate fasteners and adhesives. Your building supplier or contractor can review which materials will work best for your application. If rigid board insulation is used, make sure that all joints are tight and well taped to reduce air leakage.

3. Rigid board or batt insulation can be added to a new wall framed on top of the existing siding structure. Batt insulation requires a thicker wall to achieve the same insulation value as rigid materials.

    a. Frame the desired wall thickness on top of the existing wall siding.

    b. Add the insulation, ensuring complete coverage between studs.

    c. Apply an air barrier of spun-bonded olefin, using well-taped edges and seams.

    d. Ensure that water runoff from the roof will not drip directly onto the top of the new wall extension.

    e. Make sure that wind cannot enter the wall cavity from the top or bottom framing plates.

Upgrading Insulation on the Interior

Determine the best application method by assessing your specific requirements. Your choices include:

- Upgrading an existing wall. This is often done when the existing wall material is damaged or lath and plaster is falling off.
- Adding rigid board insulation directly to an existing wall surface. Once the wallboard is added, a new drywall finish surface is commonly added.
- Building a new frame wall. The new wall takes more space from the interior, but provides lovely window boxes. It also allows you to increase home insulation values and apply a proper vapor barrier.

Upgrading an Existing Wall

A common upgrade for older homes is removing the lath and plaster wall (a messy job at best) and replacing it with modern drywall:

- With the internal wall studs exposed, you can easily upgrade wiring, plumbing, ductwork, central vacuum pipes, and, of course, your insulation.
- Use additional horizontal strapping as shown in Figure 3-27 to increase the wall-cavity depth from a typical 2 x 4" to 2 x 6".
- Electrical boxes and window frames must be extended to allow for the increased wall-cavity depth. Building supply stores carry box extenders for such work.
- Use batt insulation layered vertically for the existing wall and horizontally for the new wall section. Alternatively, blow in cellulose insulation after the vapor barrier has been installed.

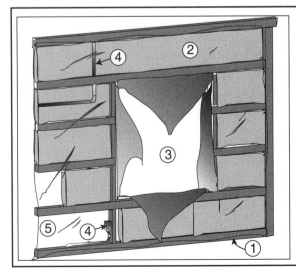

① New framing wall
② Insulation
③ Rough window opening
④ Electrical/plumbing runs
⑤ Vapor barrier

Figure 2-31 Adding a new, non-loadbearing wall will allow you to achieve increased home-insulation values.

- Install a 6-mil vapor barrier over the new framing studs.
- Apply the drywall or surface finish.

Adding Rigid Board Insulation to an Existing Wall

Rigid board insulation may be applied directly to an existing wall surface:

- As with upgrading an existing wall, windows and doorframes must be extended and electrical box extenders added.
- Your building supply store will be able to provide suitable fasteners to hold the rigid board insulation to the existing wall surface.
- Vapor and air barriers are not required when rigid board insulation is installed provided the seams are snug and well taped.
- Drywall or other finish materials may be applied directly to the rigid board insulation.

Building a New Frame Wall

You can build a new frame wall by using exactly the same techniques described earlier in *Upgrading an Existing Wall* and illustrated in Figure 2-4.

Insulating Both Sides of an Exterior Wall

A final word is required on adding insulation to both the inside and the exterior of the home. If you wish to add some of the insulation on one side of the wall and additional insulation on the other, be careful to limit the ratio to two-thirds inside and one-third outside. This mix is required to ensure that condensation does not damage the insulation system during the heating season.

Step 6 – Upgrading Attics

Air Leakage (a quick review)

Earlier, in *Step 3 – Stopping Air Leakage*, we discussed the need to prevent air from leaking into the attic space from the warm areas below. Eliminating air leakage stops the warm, moist air from condensing when it reaches colder winter air inside the attic. Condensing moisture may cause mold and structural rot. In minor cases, moisture buildup greatly reduces insulation effectiveness.

A proper vapor and air barrier **MUST** be installed on the warm side of the insulation, NOT ON THE COLD SIDE! From a practical point of view, adding a large polyethylene film to the attic-floor of the insulation is very difficult. It requires moving large amounts of existing insulation and installing the barrier around obstructions such as duct work, plumbing pipes, and wiring.

Should the interior ceiling of the house also be under renovation, you have the opportunity of adding the air and vapor barrier directly to the attic-framing members, under the drywall.

In the majority of cases, it is just not possible to add an effective air and vapor barrier into a retrofit attic. In these cases, a well-sealed and caulked attic must suffice.

Attic Ventilation (a reminder)

Attic ventilation is mandatory. A home will typically have a row of vents under the eaves called soffit vents (see Figure 2-32). These vents form the air intake for the attic ventilation system. Roof vents are installed along the roof peak (see Figure 2-33) or at the gable ends and allow hot attic air to exit. Air exiting the roof or gable vents causes suction within the attic, drawing in cooler air through the soffit vents. You will recall that a ratio of 1 unit of roof vent area to 300 units of attic floor area is required (approximately 3 square feet of roof vent space for every 900 square feet of attic floor).

Figure 2-32 Soffit vents form the air-intake section of the attic ventilation system.

Figure 2-33 Roof vents along the roof peak vent hot air out. A minimum of 1 square foot of roof venting is required for every 300 square feet of attic floor space.

Before starting any insulation work in the attic, ensure that the vents described above are present and of adequate size. If in doubt, ask your building contractor to assess their effectiveness.

Installing Attic Insulation

Once the attic has been well sealed against air leakage and the soffit and roof vents checked, you are ready to begin insulating. You have a number of choices as to which material to use. Rigid board insulation is almost never used, due to the difficulty of working in confined spaces. Blown or poured fiberglass

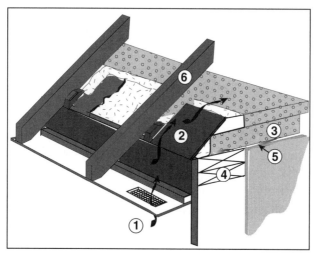

1. Soffit vent
2. Cardboard baffles
3. Insulation (R-40/RSI 7.1)
4. Top of wall
5. Vapor barrier
6. Roof truss or rafter

Figure 2-34 A well-sealed attic with plenty of ventilation and insulation is the final step in keeping your house warm in winter and cool in summer.

and cellulose are perhaps the easiest insulation materials to use. Most building supply stores will even provide a blower unit free of charge when you purchase the insulation from them. However, many homeowners simply carry bags of material to the attic, pour it, and rake it flat.

Adding insulation to the attic is not difficult. The major concern is to ensure that attic vents are not plugged with the stuff. Figure 2-34 illustrates how cardboard baffles can be stapled or pressed into the space between the rafters or joists. These baffles prevent insulation from rolling down into the soffit vent area.

Ensure that insulation is not applied around electrical light fixtures or bathroom vent fans that are not specifically approved for insulation coverage. Overheating light fixtures are a sure way to start a house fire.

Figure 2-35 This eighty-year-old home boasts the world's largest icicle, indicating severe heat loss. A professional assessment of this home demonstrated to the owners how to reduce their heating bill by at least one-third while at the same time making their home more comfortable.

2.3
Appliance Selection

What's All the Fuss About?

A home without major appliances, computers, and entertainment systems is a lifestyle that few people would embrace. There is nothing fundamentally wrong with having air conditioning or a large-screen TV (other than watching it 18 hours a day). The use of energy to power these devices is not in itself a problem. The difficulty occurs when homeowners purchase inefficient models and use polluting, non-renewable resources to power them. There is no free lunch. At some point in the future (some would say right now), these non-renewable resources will start giving up (remember the blackout of 2003?) or kill us with their exhaust.

The good news is that governments and appliance manufacturers are beginning to understand these issues. Manufacturers are actively working to lower the energy requirements of their products. For renewable-energy-system users, these improvements have exponentially increased the number of appliances available to them. For energy conservers on the grid, these same appliances are helping put money in the bank and enabling them to live just a little bit lighter on the planet.

The average 25-year-old refrigerator (the one keeping a six-pack of beer cold in the basement) uses approximately 2,200 kilowatt-hours of energy per year to operate. Electrical rates vary greatly around North America, but the average daily rate in California (when there

Figure 2-36 A standard Sears 18.5 cubic foot, 2-door refrigerator like this one will save thousands of dollars and kilowatt-hours of energy over its twenty-plus years of operating life.

is power) is approximately 15 cents per kilowatt-hour or $330 per year oper-ating cost [1]. A new Sears 18.5 cubic foot, 2-door unit uses 435 kilowatt-hours per year or $65.25 to operate, at a savings of $265 per year. At that rate, it would only take three or four years to pay for itself (assuming that rates don't climb any higher). Put another way, if you converted the savings into beer you could probably supply suds for the whole block on the savings alone! If that isn't incentive to switch, what is?

How to Select Energy-Efficient Appliances

It's not necessary to carry around an energy meter like the one shown in Figure 2-37 to measure the energy consumption of each appliance you wish to purchase, although it wouldn't hurt.

These meters (which are available from many sources, including www.theenergyalternative.com) plug into a wall outlet, and the appliance plugs into a receptacle on the front of the meter. The meter display indicates the electrical power consumed. Remember that power (watts or W) is a measure of the flow of electric-ity (amps or A) multiplied by the pressure of electricity (volts or V). Most major appliances such as washing machines, re-frigerators, dishwashers, and food processors plug into a standard 120 V outlet (pressure). The flow of electricity in amps multiplied by the pressure (volts) results in the wattage. The calculation for a typical food processor drawing 2.4 amps is:

Figure 2-37 An electronic energy meter such as this one will tell you how much power an appliance requires to operate. Most models can even be programmed with your utility rates to tell you cost over a period of time.

$$120\ V \times 2.4\ A = 288\ W$$

Compare this with the power used by an electric kettle drawing 12.5 A:

$$120\ V \times 12.5\ A = 1,500\ W$$

What does this mean? Let's suppose that you and your partner are shop-ping for a new television set. You compare all of the models and find two

---

1. Rates fluctuate from a low of 3.3 cents per kilowatt-hour in Ontario to peak daytime rates in California of 30 cents (all prices in US$). Check your utility rate and delivery charge (per kilowatt-hour) and multiply it by the yearly consumption rating of your current and proposed appliance models to calculate the savings.

that are about equal on your list of requirements. You whip out the power meter, or, if you are more conservative, authoritatively inspect the electrical ratings label, and find that model "A" uses 162 W of power while model "B" requires 1.9 A.

At this point most people would run screaming out the door having flashbacks to those grade nine "A car is moving west…"-type problems. You, on the other hand, have studied *Chapter 1.3 – What is Energy?* and recall that power (in watts) is the voltage multiplied by the current.

120 V x 1.9 A = 228 W

You close the deal by smoothly informing the salesperson that model "A" is the better TV as it will save you loads of dough over its operating life, reduce greenhouse gas emissions, or require less electrical-generation equipment if you are using off-grid energy sources. You might even save money on the purchase with the salesperson trying to get rid of you.

But power is only half of the equation. If for some strange reason you were to watch model "A" television 2 hours per day and model "B" for 1-1/4 hours per day, the energy consumption for model "B" would be lower:

Model "A" energy consumption = 162 W x 2 hours = 324 W-hours
Model "B" energy consumption = 228 W x 1.25 hours = 285 W-hours

Time factors into the energy cost. This is why you should turn off lights in empty rooms.

There Has to Be a Better Way

Most people are not so fanatical as to carry an energy meter with them when they go shopping for appliances. Some people may take a look at the electrical ratings label, provided they can keep the watts and volts straight without having to carry a copy of *$mart Power*. For the rest of us, the government has made life a little easier, at least for the larger appliances. Figure 2-38 shows a Canadian "EnerGuide" label that is affixed to a high-efficiency washing machine. The EPA uses a similar label for its "EnergyGuide" program. Both programs require that labels be affixed to appliances, providing comparison data for appliance models of similar size with similar features and indicating the appliance's total energy consumption per year.

A closer look at an EnerGuide label reveals the following features:

- The bar graph running from left to right represents the energy consumption of all models of similar appliances.
- At the left side of the graph is the energy consumption of the most efficient appliance in kilowatt-hours per year.

• At the right side of the graph is the energy consumption of the least energy-efficient appliance in the same class.

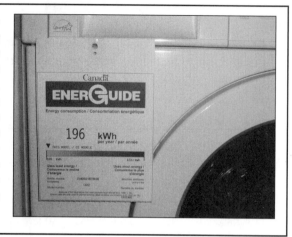

Figure 2-38 This high-efficiency washing machine carries both the "EnerGuide" and "Energy Star" program logos, allowing consumers to quickly compare the energy consumption of different models. The Energy Star logo indicates to the consumer the appliance meets minimum energy efficiency ratings.

The bar graph on the EnerGuide label indicates energy consumption of 189 kWh per year for the most efficient appliance and 1032 kWh per year for the least efficient. Think about that for a moment. Two appliances of the same size and class, one consuming almost five-and-a-half times the amount of energy to do the same job!

Efficient washing machine = 189 kWh per year x 10 cents per kWh = $19 per yr.
Inefficient washing machine = 1,032 kWh per year x 10 cent per kWh = $103 per yr.

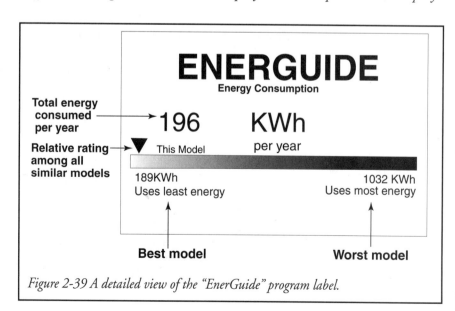

Figure 2-39 A detailed view of the "EnerGuide" program label.

Figure 2-40 This Sears high-efficiency clothes washer uses five-and-a-half times less energy than a similar-sized unit. Also consider the savings on soap and reduced waste water being pumped into the sewer or septic tank.

Assuming a delivered price of 10 cents per kilowatt-hour of electricity, the difference in operating expense is $84 per year. Over the life of the machine (say twenty years), that's a whopping $1,700!

In Figure 2-39, the triangle-shaped pointer over the bar graph on the EnerGuide label shows the energy consumption of this particular model in relation to that of the most and least efficient models in the same class. The closer the pointer is to the left side of the bar graph, the lower the operating costs.

The sample label used in Figure 2-39 is from a Sears front-loading washing machine, straight from the

Figure 2-41 Not only is the solar-powered clothes dryer energy efficient; clothes also last longer, smell better, and don't require yet another artificial fragrance sheet.

catalog pages. The most efficient model is the Staber horizontal-axis machine. The Staber is almost twice as expensive as the Sears model but only 1% more efficient. On the other hand, the least efficient model is marginally more expensive than the Sears high-efficiency model.

Small Appliances

The EnerGuide program was designed to take care of major "white goods" appliances. What about smaller appliances and electronics that don't carry the program label?

Our choices at this point become a little bit more difficult. We have to revert to using an energy meter or reading the electrical-ratings label on the device. An electrical meter such as the one shown in Figure 2-37 is a great device provided that you have access to the appliance for a sufficient amount of time to conduct a test analysis. This is important because energy consumption requires time to evaluate (energy = power multiplied by time).

Let's take a look at a conventional coffee maker to see how time factors into our assessment. A typical coffee maker such as the one shown in Figure 2-42 has an electrical rating label showing 120 V and 10.5 A. Whipping out your calculator you correctly arrive at a power consumption of 1,260 W.

Figure 2-42 A coffee maker uses a lot of energy to boil water, but considerably less to keep the pot warm. In the absence of an EnerGuide or EnergyGuide label, use care when calculating energy consumption based on the manufacturer's label.

120 V x 10.5 A = 1,260 W of power

At 7:00 o'clock on Saturday morning you stumble down the stairs and get the brew going. By 11:30 you have slugged back your third cup, draining the machine and shutting it off. Applying your caffeine-honed mathematical skills, you determine that the coffee maker was on for 4 1/2 hours, giving you an energy calculation of 5,670W.

1,260 W of power x 4.5 hours = 5,670 watt-hours of energy or 5.7 kWh

You realize that at 10 cents per kilowatt-hour for energy your morning coffee has just cost you 57 cents worth of energy, right?

5.7 kWh x $0.10 per kWh = $0.57

Wrong. The math is correct, but the assumptions are wrong. You have to be very careful when calculating the energy consumption of an appliance, as it may in fact change with time. Yes, the coffee maker label does say that it requires 10.5 A or 1,260 W of power. What it does not tell you is that it only needs that much power to boil the water. Once the coffee is brewed, it uses less power to keep the pot warm.

So what is the correct answer? Actually, I don't have a clue. Without access to the coffee pot and an energy meter, it's anybody's guess. When trying to calculate the energy consumption of any small appliance with no electrical rating label, follow these general guidelines:

- Appliances that draw a lot of wattage, typically over 300 W, should not be used unless it is only for a short period of time. This applies to coffee makers, hair dryers, curling irons, electric kettles, clothes irons, car block heaters, and space heaters. Crock pots which slowly simmer food for hours at a time are actually more efficient than ovens.

- Coffee makers such as the model described above can be used, but transfer the fresh coffee to a thermos. The coffee stays warm and tastes better without using unnecessary energy.

- Consider boiling water using an electric kettle (instead of the stove) and a drip coffee basket. Transfer the fresh coffee to a thermos to keep it warm.

- Refer to *Consumer Guide to Home Energy Savings* for electrical consumption ratings of desired appliance models before you buy. (The book is available through Real Goods (www.realgoods.com), part number 82-399.)

- Refer to Appendix 2 of *$mart Power* for a list of various electrical appliances and tools with their power and electrical ratings.

- All major heating appliances such as cook stoves, ovens, electric water heaters, electric clothes dryers, furnaces, central air conditioners, etc. draw an enormous amount of electrical energy. Always purchase the most efficient model and consider the addition of timers or other appropriate controls to regulate their operating time.

- Microwave ovens are much more efficient than regular electric models. This is an especially important consideration in the summertime when waste heat from the oven must be removed with expensive air conditioning.

- Try to purchase appliances that are not equipped with an electronic clock or "instant-on" anything. These devices are considered phantom loads and consume a large amount of power without doing anything for you. Cell phones, PDAs, MP3 devices, and other electronic items that require battery charging can be plugged into a power bar. Turn the power bar off once the device has been fully charged.

- Convert as much lighting as possible to high-efficiency compact fluorescent lamps (see below).

- Use high-efficiency front-loading washing machines. The Staber HXW-2304, for example, is so efficient that you require just one ounce of soap per wash load. You will use less electrical energy to operate the washer, less water per cycle, and less heating energy to heat the smaller amount of wash water.

- If you are not sure about an appliance or tool, borrow one from a friend and plug it into an energy meter like the one shown in Figure 2-37. Even if the appliance model is not exactly the same, this will give you a general idea of the power and energy consumption.

Computers and Home-Office Equipment

Many people are finding that self-employment is the way of the future, especially after the last round of corporate downsizing. An essential part of working and playing at home now revolves around computers and related home-office equipment.

With your computer, desk lamp, printer, modem, and monitor working overtime (maybe all the time if you have teenagers), how will your electrical energy consumption be affected?

Modern electronics are a marvel of efficiency and they keep getting better all the time. For example, let's compare a 7-year-old, 15-inch color computer monitor with a new flat-screen model. The older unit may use as much as 120 W of power, while the new one draws about 30 W. That's a 75% reduction in power usage.

You should also consider that few people use a computer for only a minute or two and then turn it off. The average user may have the computer and support system operating for many hours, which increases energy consumption (power x time). It makes sense to purchase the most efficient products because of this longer operating time.

Don't just think about the computer. You will almost certainly have a printer, monitor, desk lamp, room lights, fax machine, and possibly a photocopier to fully equip your home office. Teenagers (and aging rockers) can add to this list a 400 Watt surround-sound system. Running an energy-effi-

Figure 2-43 A wall full of home-theater equipment may not seem very energy efficient, but in fact it is. Modern electronics are a wonder of efficiency and work well with any renewable energy system. Just remember to switch off the power input using a special plug or power bar to eliminate phantom loads.

cient system for forty hours a week will still burn up a lot of juice.

Fortunately, the government has come to our rescue once more. The Energy Star program has been adopted across North America and is the equivalent of EnerGuide and EnergyGuide labeling. Energy Star provides guidelines and requirements to manufacturers who seek compliance for their electronic stereo, TV or data-processing products. Most compliant products use high-efficiency electronic components and specialized power-saving software. For example, a compliant monitor will enter a low-power sleep mode if the image has not changed within a given time frame. Another example is a laser printer which adjusts its heater temperature when idle. A laptop computer will always be more energy efficient than the equivalent desktop model, owing to the limited energy stored in the batteries. Likewise, a bubble jet printer is more efficient than a laser printer.

As with everything else in life, there are choices to be made. If you want lower per-print costs, then laser-printer toner is cheaper than bubble-jet cartridges. Laptops are more expensive than a desktop system. Just remember to shop the energy labels and look for the Energy Star logo.

Figure 2-44 Purchasing computers and home-office equipment with the Energy Star label ensures the lowest energy usage.

2.4
Energy-Efficient Lighting

Possibly the single most important invention to touch our lives is the incandescent light bulb. Prior to Mr. Edison's 1879 discovery, it was almost impossible to stay up past sundown. (My wife still has this problem.) Life before the modern light bulb meant filling a kerosene lamp and enduring poor lighting and air quality. In the early days of electrical-power production, little thought was given to energy efficiency or environmental concerns. As a result, light bulbs were—and are—inefficient and wasteful. The bulb in your floor lamp has remained essentially unchanged for over a hundred years. However, times have changed and so have lighting technologies.

The incandescent lamp and its many variations create light by heating a small coil or filament of wire inside a glass bulb. Anyone who has watched a welder working with metal knows that hot metal glows a dull red color. If the metal is heated further, it glows brighter and with an increasingly whiter light. At some point, the

Figure 2-45 We owe a nod of thanks to Mr. Edison and an entire industry that developed a replacement for the kerosene lamp. As technology marches forward, however, it is time to say goodbye to our old but wasteful friend—the incandescent lamp.

metal will vaporize with a brilliant shower of white sparks. The various stages of glowing metal are known as incandescence.

In a similar manner, electrical power applied to a light bulb causes the metal filament to glow white-hot. If you decrease the amount of power reaching the bulb (by using a dimmer switch, for example) the bulb dims and glows with a progressively redder color.

Making an incandescent lamp glow requires a large amount of energy to heat the filament. In a typical light bulb, 90% of the energy applied to the filament is wasted in the form of heat. Therefore, only 10% of the energy you are paying for is making light. What a waste!

To make matters worse, we need to think about the heat component for a moment. If it is summertime, this waste heat contributes to warming the house and increasing the air-conditioner load. An incandescent bulb hits you twice in the wallet: poor efficiency resulting in high operating cost and waste heat as well as the expense of getting rid of the waste heat using air conditioning.

Let's put this into perspective. Assume it's a warm summer evening. You have a total of fifteen 100 Watt lights on in the house. The waste heat is:

15 bulbs x 100 W x 90% heat loss = 1,350 W of heat

This is about the same amount of heat output as you get from a large electrical space heater in the winter.

You may argue that this waste heat can be used in the wintertime to help warm the house. True, except that electricity is the most expensive means of heating a home anywhere in North America and this byproduct of lighting cannot be controlled unless you want to place your house lights on a thermostat. If you are generating your own electricity off-grid, this waste is not even manageable. So what are the alternatives and how do they save energy and money?

There are numerous alternative lighting technologies, including semiconductor-LED, mercury vapor, halogen, fluorescent, halide, and sodium. These technologies have a range of applications and different efficiencies. The most common and efficient technologies for home applications are discussed below.

Before we continue it would be prudent to say a few words about semiconductor or LED lighting. Most people have seen LED lights installed in automotive applications or high-efficiency flashlights. Currently, LED lighting is only one-half as efficient as compact fluorescent technology. Additionally, the color quality of these bulbs is very poor and would not be acceptable for home lighting. Long life and low energy consumption compared with standard incandescent lamps gives LED lights their passing grade.

Compact Fluorescent Lamps

Mention fluorescent lamps to my wife and all she can think of are the pasty-faced girls applying makeup in the washroom at the high school prom. In the 60s and 70s some companies used the term "cool white" on their fluorescent lamps, as if this were a good thing.

Enter the compact fluorescent lamp (CF). These marvels of efficiency may look a little odd, yet they offer many advantages over standard incandescent lamps. A CF lamp is designed to last ten times longer than an incandescent lamp. Shop around when looking for these lamps. Stores such as

Wal-Mart carry the electronic CF lamps for about $3.00 each. Many stores still treat CF lamps as "special," which includes a "special" price of about four times this amount.

Light output and energy efficiency are further considerations. Contrary to popular belief, light output is not measured in watts, but in *lumens*. The wattage rating of a bulb is the amount of electrical power required to make it operate. The standard 75 Watt light bulb gives off approximately 1,200 lumens of light. A 20 Watt CF lamp provides the same intensity but uses one-quarter the energy. Translate this to cost savings and the CF bulb will save $55.00 over its life span, assuming delivered energy costs of 10 cents per kilowatt-hour. This does not even take into account the savings in air-conditioning load and environmental pollution.

Figure 2-46 The compact fluorescent (CF) lamp is the bulb of choice in modern energy-efficient homes. A standard 60 Watt incandescent lamp is shown at the left. An equivalently bright 15 Watt CF lamp is shown immediately to its the right.

Lastly, CF lamps won't give you headaches or the Bride of Frankenstein look first thing in the morning. Advances in phosphor coatings and electronic ballasting eliminate ghastly color and flicker. The lighting industry applies a rating system called the Color Rendering Index (CRI) to all bulbs, although it can be a bit hard to locate the information. The closer a light source's CRI rating is to 100, the more natural and comfortable it will be to the eyes. The typical incandescent lamp has a CRI rating of 90 to 95 compared with a CF-lamp rating of 82. Compare this with a Frankenstein fluorescent bulb rating of 51. Still not convinced? Buy a couple of CF bulbs and try them out. It will be almost impossible to tell the difference in color or operation when they're in operation.

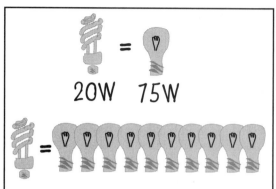

The Down Side

The shape and size of many CF lamps are not identical to that of standard bulbs. This may cause problems fitting the CF lamp into a conventional socket. Take note that shapes and sizes vary by manufacturer. Check sev-

Figure 2-47 CF lamps last ten times longer than regular light bulbs and use approximately four times less energy. With no flicker and color quality similar to incandescent light bulbs, you won't get headaches or see that awful pasty color when you look in the mirror.

eral different brands to see if one will fit your application. Before you resort to changing the light fixture, see if your local hardware store is able to suggest an alternate base, harp, or socket extension.

Dining rooms often have overhead fixtures that are connected to a dimmer switch. CF lamps must never be placed in these sockets unless they are specifically designed for dimming. Dimmable CF lamps are available, but at a premium price.

CF lamps can be used in outdoor lighting systems even in winter. In this application, the lamps typically require one to two minutes to warm up and produce useful levels of light. This is not a problem for security or perimeter lighting on timers. Garage-door-opener lights or those applications where the light is required quickly should not use CF lamps.

T8 Fluorescent Lamps

Large-area or kitchen-cabinet indirect lighting often uses standard fluorescent lamps with magnetic ballasts. Although more efficient than incandescent lamps, these models are still only half as efficient as CF lamps or new "T8" fluorescent lamps with electronic ballasts. T8 lamps are more expensive than CF units but they are excellent replacements for older 4' and 8' standard fluorescent tubes. It's easy to recognize T8 lamps, as they are approximately half the diameter of conventional tubes. Likewise, electronic ballasts are very light for their size, especially when compared with older, less efficient magnetic ballasts.

Track Lighting and Specialized Lighting

There are places where even a die-hard energy conserver would never put a CF lamp. Lighting your Rembrandt or Picasso collection is one area that comes to mind. For these applications, low-voltage, high-intensity MR-16 lamps are ideal. Jewelry and watch stores use these miniature spotlights for clear, intense lighting of small objects, artwork, and display cabinets.

MR-16 halogen gas incandescent lamps are fabricated with specialized internal reflectors that direct light in a unidirectional flood pattern. MR-16 bulbs operate on 12 V, which in turn requires a transformer (supplied with the light fixture) to drop the household 120 V supply. The low wattage of MR-16 lamps is similar to that of a larger CF lamp, so they can be considered efficient in these special applications.

Figure 2-48 MR-16 flood lamps are excellent for lighting your priceless jewelry or baseball cap collection. Although they are incandescent lamps, their low wattage and high intensity make them acceptable for these applications.

Large-Area Exterior Lighting

Large outdoor areas require major lighting muscle. While it is possible to use a few million CF lamps to light your yard, an easier and even more energy-efficient method is to install high- or low-pressure sodium lamps. These have a characteristic yellow color that makes them suitable for perimeter and security lighting. They also work well in horse-riding arenas and for street lighting. Watt-for-watt, a sodium lamp produces twice as much light with the same power as a CF lamp and is up to ten times more efficient than an incandescent lamp.

Free Light

Perhaps the best lighting is free—100% energy efficient. No, don't just open the drapes. Even the best-designed house may have an area or two that suffer from low light levels during the daytime and require lighting even on the sunniest days. If you're thinking skylights, remember that typical skylights are not energy efficient, are prone to condensation problems, and often require reframing and finishing from the inside. A product available from www.sunpipe.com offers an excellent alternative. The Sunpipe is similar to a fiber-optic device on steroids. An intake piece is mounted on the roof in the same fashion as the metal chimney shown in Figure 2-49a. A reflective supply pipe is fitted down through the roof opening into the living area. A trim kit is installed and that's it.

Figure 2-49a This view details the "intake" portion of the Sunpipe unit.

The Sunpipe directs outside light into the immediate area, even on cloudy days. An example installation is shown in Figure 2-49b with the Sunpipe illuminating a narrow second-floor hallway. If the Sunpipe were to be covered, this hallway would be almost completely dark. A 9"-diameter pipe provides illumination equivalent to that of a 400 Watt incandescent lamp on a sunny day.

Figure 2-49b The hallway in this picture is completely illuminated with a Sunpipe. Covering the intake to show the difference without the light from the Sunpipe created an almost-black photograph.

2.5
Water Supply and Energy Conservation

Introduction

Turning on a tap for a glass of water is a luxury that few people give any thought to. If you live in the city, all it takes is writing a check to your local utility a few times a year and presto! Clean, clear, life-sustaining water.

Society's lack of concern for or ignorance about this natural resource is often beyond comprehension. One only has to drop by a golf course in the desert to see firsthand the waste of a precious resource. Many people counter with: "Why not use water as we wish? Isn't water a renewable resource, like the solar energy we are capturing?"

Water is a plentiful and renewable resource, within limits. The majority of the earth's water is stored as moisture in the air, salty seawater, or ice and snow. The rest is in lakes and streams or in wells drilled to reach the underground water table.

Figure 2-50 An average person in North America consumes 100 gallons (382 liters) of water per day. The conserver can easily reduce this consumption by 50%.

Lake- and ground-water levels have been dropping for many years; hardly a day goes by without some newspaper headline screaming about the subject. Consider that the Colorado River does not always flow into the ocean, partly because of over-extraction in the upstream watershed. The reduction in water levels is largely the result of increased human population and water usage. Industrial and climatic changes are also being blamed for an ever-increasing reduction in our second most important natural resource after air.

The average person in North America consumes 100 gallons (379 liters) of water per day. If we include the water required for industrial and agricultural processing, consumption rises rapidly to over 1200 gallons (4543 liters) per day. North Americans consume water at a rate 5-1/2 times greater than that of people in Sweden. As with gasoline, our wasteful habits can be traced back to costs. A 1988 survey of OECD countries showed that efficiency and price were interrelated, with Germany charging $2.10 per cubic meter of water and Canada practically giving it away at 31 cents.

Besides saving money on water bills and energy bills for hot water, reducing our consumption will lighten the load (electrically and literally) on water- and sewage-treatment facilities. This is a significant consideration when you factor in upgrades in capacity to aging infrastructure that would almost certainly be adequate if conservation of water and sewage were undertaken. The savings to these municipalities (and property taxpayers like you and me) would be enormous.

Water is also very heavy, requiring large amounts of energy for the operation of the municipal supply and sewage pumps.

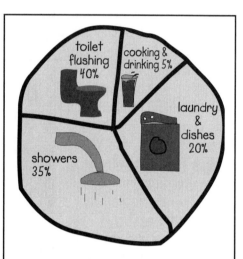

Figure 2-51 *This pie chart shows how the average North American consumes water. Lawn and garden water is subject to high variability between homeowners and apartment dwellers and has been averaged into each section of the chart.*

The Conserver Approach

It is estimated that 75% of our water consumption occurs in the bathroom and 20% for laundry and dishes. The remaining 5% is used for cooking and drinking. Conserving water and the energy used to heat and move it requires the same approach as with lighting and appliances: do more with less. This does not mean that the whole family has to get in the shower at the same time. It means getting the same results with more efficient appliances and methods.

Toilet Flushing

Conserving water when flushing a toilet is not only simple, it is actually the law in many areas wracked with inadequate water supplies. Simply remove the old clunker and replace it with a certified low-flush model. This conversion will reduce water usage by 66% or more. Conventional toilets can consume over 4.7 gallons (18 liters) per flush while low-flush models use only 1.6 gallons (6 liters).

Low-flush toilet models are available in a mind-numbing array of styles, colors, and shapes to suit even the most discriminating derriere. Prices keep coming down, with economy models available in the $50 range. Another good reason to consider these models is the lack of condensation dripping from the tank. If your old clunker drips every time the humidity rises, that's the excuse you need to make the swap.

No matter which brand of toilet you own, if it leaks it's a water waster. A toilet that leaks four gallons of water per hour

Figure 2-52 Low-flush toilets require only 33% of the water required by earlier models and come in a variety of styles to suit even the most discerning derriere.

(fifteen liters) would fill a large swimming pool if allowed to continue for a year. Not sure if the toilet leaks? Place a couple of drops of food coloring in the water reservoir. Check the bowl after fifteen or twenty minutes and see if the color shows up. If it does, the next book you need is *An Introduction to Plumbing* to fix it.

Bathtubs

If you and your family prefer baths rather than showers, there is not much you can do except change an older bathtub to a new fiberglass model. Old cast iron or metal tubs that are six feet long and very deep require a lot of water and a large amount of fuel to heat it.

A fiberglass soaker bath can be ordered in a 5-foot model that is fine as long as you aren't related to Magic Johnson. A shorter bath requires less water volume and the fiberglass model resists heat loss through the sides.

If you are having a bath during the heating season, leave the water in the tub until it cools down, and then pull the plug. This little trick will transfer the heat energy from the water into the room, providing you with free warmth and humidity to boot.

Showers

Aside from turning off the water when the teenagers decide to live in the shower, the best solution is to add a low-flow showerhead. These units, such as the one shown in Figure 2-53, have built-in water-flow restrictions and special "needle orifices" to reduce flow without making you feel as if you're just standing under a leaky roof. Look for models that carry a label from a certification agency displaying a flow rate of less than 2.6 gallons (10 liters) per minute. Many models come equipped with a built-in massage feature, providing an additional incentive for making the switch.

The Kitchen

A very quick upgrade of the kitchen sink may be in order. Try replacing a standard-flow faucet for one with an "aerator" noz-

Figure 2-53 Showerheads such as this model reduce water consumption and hot water heating costs by 50%.

zle, such as the one shown in Figure 2-50. These devices mix air with the water stream, giving the appearance and feel of more water flow. This apparent increase in water flow offers more coverage when washing dishes and hands.

Dishwashers

Like every other major appliance in the house, dishwashers are subject to the government's EnerGuide and EnergyGuide programs. When purchasing a new model or updating a tired machine, review the EnerGuide or EnergyGuide label and purchase the most efficient model available. Better yet, try washing your dishes by hand. The energy savings provided by "manual" dishwashers is quite high, and you might find this a good time to catch up on family gossip.

Clothes Washers

Washing machines are in the same boat as dishwashers. Check the EnerGuide or EnergyGuide labels and purchase the most efficient machine you can afford. The washing machine shown in Figure 2-40 consumes five-and-a-half times less energy than a similar new machine. Shop carefully and save big!

Water Heaters

Besides changing your water heater completely, there are a few things that can be done to help conserve energy:

- Turn down the thermostat of the water heater to the lowest temperature you can accept. The thermostat on most gas units is visible and well marked. Electric units often have two thermo-stats underneath a re-movable cover. When adjusting them, ensure that the power to the unit is turned off. If you are unsure, contact your plumber for assistance.

- Install an insulation blanket as shown in Fig-ure 2-54. These blankets are available from most building supply stores or plumbing contractors. The insulation batting helps to stop heat trans-fer from the hot water inside the unit. If you create a homemade blanket, make sure it is fireproof and that it does not block any air intakes or vents on a gas water heater.

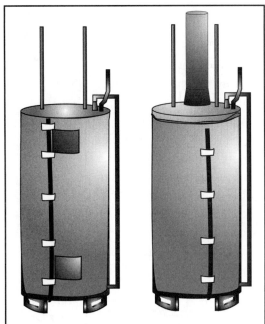

Figure 2-54 Water heater blankets save a great deal of the heat that is typically lost from hot water storage tanks. Make sure the blanket does not block any air intakes or vents on gas water heaters.

- Flush the water heater at least twice per year. Built-up sediment in the bottom of the tank reduces heater efficiency. Draining a few gallons from the bottom of the tank will reduce calcium and mineral buildup.

- Install pipe insulation on all hot water lines. This material is inexpensive and very easy to install (provided that the pipes are exposed).

- Add an active solar heating system that captures the sun's rays to warm the water in your tank. Solar thermal systems can easily provide 50% of your annual hot water heating requirement and provide financial re-turns on investment of 10% or better. Chapter 3 covers this equipment in detail.

Water heaters have an average life span of ten years. Before you discover a new indoor swimming pool courtesy of your leaking unit, install an upgraded model. A water heater that is reaching the end of its useful life is a bit like a refrigerator: the savings resulting from the energy efficiency of newer models will easily offset the installation cost.

New to North America are the instantaneous water heaters such as the Bosch model shown in Figure 2-55. These units have long been used in Europe where energy costs prohibit wastefulness. Your plumber can easily replace your old storage unit and give you back a bit of floor space at the same time. Do not be misled by the old wives' tale that on-demand water heaters cannot supply sufficient hot water for the entire house. This is a fallacy of days gone by which is propagated by ignorance of the technology.

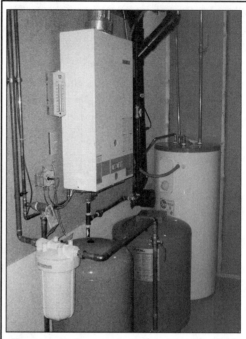

Figure 2-55 Instant water heaters such as this model from Bosch do not waste energy by storing hot water all day long. They work by rapidly heating the incoming cold water, supplying enough hot water for the entire house. Do not be misled by the old wives' tale that on-demand water heaters cannot supply sufficient hot water for the entire house. This is a fallacy of days gone by which is propagated by ignorance of the technology.

Solar Hot Water Heating

Solar water heating systems are a natural extension of efficient water heating systems and are discussed in Chapter 3.

Landscaping

Lawns

Possibly the greatest waste of water (and energy) is using it to maintain the perfect lawn. If there was ever a sinkhole for energy, this is it. We start by trucking in tons of topsoil and adding fertilizers, pesticides, and lime to get it growing. Then we use more fuel to cut, trim, and edge our lawns, all the while wasting energy. Added to this is our need to use purified drinking water when rain would almost certainly suffice. Why water in the middle of a hot summer day and have 50% of the water evaporate or run along the sidewalk? If we must have perfect lawns, then we should try to water them more efficiently.

Watering at dusk or early in the morning is best. Apply only as much water as your local conditions require. Check with a nursery or garden center to find out how much water is appropriate. Overspraying the lawn and watering the sidewalk is just plain wasteful. Spend the $20 on a better sprinkler unit and maybe a few dollars more on a timer. Program the sprinkler to suit the size and area of

Figure 2-56 An automatic sprinkler system supplied by nature in the form of rain will save a considerable amount of purified drinking water that most people use to water their lawns. Try switching to drought-resistant white clover which has the added benefit of slow growth, reducing the amount of cutting required.

your lawn. If you are planting a new lawn, consider low-maintenance ground covers or drought-resistant white clover grasses instead.

Are you in the market for a lawnmower? Consider a rechargeable electric model: virtually no noise, minimal maintenance compared to gas models, and negligible pollution. You might wish to consider a modern reel mower. Easy to use and burn calories at the same time.

Flower Gardens

The roof of your house captures a lot of rain during a downpour. Try containing some of this water in a series of rain barrels that are interconnected to supply flowerbeds. A few 50-gallon (200-liter) barrels interconnected with poly-pipe as shown in Figure 2-57 makes a nostalgic-looking storage system. Connect the rain barrels with weeping hose and you have an energy-efficient and labor-saving watering can. Remember that mulch doesn't just look good

Figure   2-57 Capturing rainwater is a time-honored tradition. Automate the process by connecting the barrels to a weeping hose for an automatic and labor-saving watering system.

and suppress weeds; it also keeps the soil from drying out, further conserving water and energy—and your time.

Although garden lighting does not help with water conservation, it helps to showcase your pretty flowers and hard work. This lighting will better harmonize with nature if it is solar powered. The price of solar-powered lights has been dropping while their quality has steadily increased. The model shown in Figure 2-58 was purchased for under $5.00 and will continue to run long after you should be in bed.

Figure 2-58 Solar-power garden lights won't help with water conservation, but they certainly accentuate the beauty of the flowers.

Water Conservation Summary

The tasks described above will easily reduce your water consumption by more than 50%. Add to this the decreased electrical energy required to pump the water, reduced water heating costs, decreased soap consumption, and diminished load on the municipal supply and sewage system, and you simply cannot justify putting off these upgrades.

Figure 2-59. Although the urban lawn is ubiquitous, it consumes an enormous amount of energy, water and your time trying while trying to make it look better than your neighbours. Xeriscaping (reducing resource requirements in landscaping) offers a beautiful and low-maintenance alternative as shown above.

2.6

A Kaleidoscope of *Eco-nomic* Ideas for Renters and Homeowners Alike

Just because you don't actually own a piece of terra firma doesn't mean that you can't or shouldn't get on the energy-efficiency bandwagon. How? Check off the items in this photo collage as you complete them. They will start you on the road to a lifelong relationship with *eco-nomics*.

☐ Replace all incandescent bulbs with compact fluorescents

☐ Let mother nature dry your clothes

☐ Use ceiling fans to cool your living space

☐ Vacuum or replace your furnace filter

☐ Program your thermostat for when you are not at home

☐ Regularly clean the lint trap in your dryer

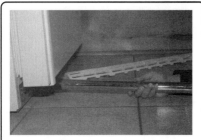

Vacuum the coils on your fridge

Add a timer and or photocell to control exterior lights.

Replace old appliances with efficent ones

Look for the Energystar logo when purchasing appliances

Replace computers with laptops and LCD displays

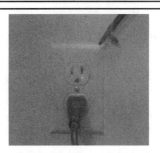

Smoke test your receptacles and install outlet seals

Use dimmer switches

Use Sola-tubes to bring natural light into your living space

Stop using heat producing halogen lamps

Use motion sensitive exterior lighting

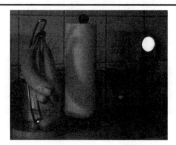

If you use nightlites, use ultra efficient electroluminescent ones.

Insulate hot water pipes

Use a rain barrel to capture rain water for garden watering

Put aerators on all faucets

Use drought resistant ground covers like clover

Use a low flow shower head

Shop locally.

Make sure all your toilets are low flush.

Get out of your car and onto your bike

Get out of your car and onto a bus

If you must use a car, fill up with ethanol

Keep tires inflated to the proper pressure

If you use a block heater, put it on a timer

Drive an economical automobile.

3
Heating and Cooling with Renewable Energy

Price signals capture everyone's attention. When oil was a couple of dollars a barrel very few people worried about how much of it they consumed to operate cars and heat homes. In the early 1970s automobiles weren't much smaller than the tanker ships used to flood North America with Saudi Arabian crude, and home design and construction gave little attention to energy efficiency and related fuel consumption.

In 1973 an organization known as OPEC unilaterally raised oil prices, providing the wake-up call that shook Americans out of their fuel consumption complacency. Prices soared and those who heated their homes with oil went from being energy pacifists to being scared silly almost overnight. As if that wasn't bad enough, all Arab oil producers stopped shipping oil to the United States as a way of protesting American policy regarding Israel. Price increases of over 500% coupled with uncertain supplies exposed the vulnerability of the United States economy, which depended then as it does now on foreign oil supplies.

Following considerable political posturing by the Nixon administration, the embargo was lifted and life returned to an uneasy state of normal. In the years following the energy crisis considerable work was done to improve the efficiency standards of homes and automobiles. Unfortunately politicians have very short memories, and the crisis of 1973 was quickly forgotten as fuel prices and supplies stabilized, causing most North Americans to return to their gluttonous ways.

The beginning of the 21st century has brought about a new energy crisis fostered not by embargoes but rather by concerns about the environment, supply, and terrorism. Although the economy is still completely addicted to oil and demand continues to rise unabated, OPEC remains relatively friendly to the West. It should; after all, American consumers have drained the astounding sum of $7 trillion from the economy simply to burn Arab oil and watch it go up in smoke. According to a report in the October 25, 2003 issue of *The Economist*, this estimate does not include subsidies of staggering cost, such as the ongoing military presence in the Middle East as well as development subsidies and tax breaks to the large oil companies.

Environmental degradation continues unabated, as neither the government nor oil companies consider it a cost to be added to the balance sheet. Canada exports nearly 8% more oil to the United States than does Saudi Arabia. The province of Alberta, through its phenomenally dirty tar sands oil extraction program, reaps the short-term economic windfall even though the process makes it one of the largest greenhouse gas emitters in the world. With governments addicted to increasing gross domestic product and job creation, energy efficiency and the environment are of no more concern than they are to the Saudi Arabian oil sheiks with whom they compete for market share.

The long-term view of energy is to move the world economy away from nonrenewable resources to clean-burning, carbon-neutral renewable fuels such as ethanol, biodiesel, and perhaps hydrogen for use in transportation infrastructure. Wind turbines, photovoltaic panels, and clean-burning biomass provide complementary energy sources for the production of electricity. The transition has already started. In 2003 Germany announced that it had already decommissioned one nuclear power plant as a result of increased use of wind turbine and photovoltaic electricity generation technologies. On the residential home front many renewable energy technologies such as solar domestic hot water systems are currently available and economically viable.

Renewable Fuels

Burning any fuel has some negative impact on the environment. Fossil fuels are nonrenewable and contribute smog-producing chemicals and greenhouse gases—mainly carbon dioxide—into the atmosphere. Some nonrenewable fuels are better than others. Natural gas and propane are the best, while low-grade coal is the worst. Consider that a natural gas or propane cook stove allows the burning fuel to vent *into* the house. Try *that* with coal or oil!

Wood and wood pellets, dried corn, biodiesel, and ethanol are different from fossil fuels in that they are renewable, easily replaced energy storage units. A growing plant absorbs nutrients and water from the ground as well as carbon dioxide from the atmosphere. Photosynthetic processes within the plant convert carbon dioxide gas into carbon, which is stored in the structure of the wood or oil seed. When dry wood or biofuels are burned, carbon is consumed, releasing heat and carbon dioxide gas. The good news is that burning the plant products properly exhausts no more greenhouse gases into the atmosphere than simply letting the tree or plant rot on the forest floor or farmer's field. However, allowing even the best wood stove to burn smoldering, smoky fires releases ash and other pollutants into the atmosphere. (See Chapter 5, "Biofuels", for more information on this subject.)

Solar Energy – Something for Everyone

If you are designing a new home or retrofitting an old one, an even better energy source to consider is the sun. A properly designed and oriented home absorbs free energy on sunny days, helping reduce your reliance on burning anything. Both passive and active solar heating systems can be used. Passive solar heating is just that: passive use of properly oriented walls, windows, and architectural house features. Active solar heating involves a series of solar collection and storage units designed to increase the "density" of the captured energy. Within the broad range of renewable fuels and "solar thermal" technologies are a number of options for the urban homeowner to consider, each with its own pros and cons. While the number of configurations is virtually infinite, let's assess those technologies that follow the *eco-nomic* path:

- 3.1 Passive solar space heating
- 3.2 Active solar air heating
- 3.3 Active solar water heating
- 3.4 Solar pool heating
- 3.5 Active solar space heating
- 3.6 Earth-energy with "Geoexchange"
- 3.7 Heating with renewable fuels
- 3.8 Space cooling systems

3.1
Passive Solar Heating

Solar energy is free, non-polluting, and renewable. Why don't more people use it? My guess is that most people think that houses using solar energy have to look like something George Jetson would live in. Well, don't worry. Geodesic dome designs aside, you won't have to live in a house that upsets your local town hall planning committee or causes you to be barred from block parties for life in order to live with a passive solar system. Another possible reason for being afraid of solar energy is its variability and the fact that your building contractor isn't able to quantify it. If you require heat, just install a big furnace and then forget about it, right? Everyone knows it's not quite that easy with solar energy.

Not so fast. Living with solar energy does not require a house full of complicated electronics and miles of glass. What passive solar energy does require is following some fairly simple guidelines to allow your home to take full advantage of the sun's energy.

Step 1 – House Orientation
Orienting the house to collect solar energy may seem obvious, but it's not. Most people give no thought to ensuring that the house is oriented in such a way as to capture as much sun as possible in the winter and provide proper shading in the summer. This is how it's done:

- Using a compass at the proposed building site, locate magnetic south. Consult the magnetic declination chart in Appendix 4 and determine the compass correction for true or solar north and south. Using this heading, place the long axis of the home at a right angle to it.
- Place deciduous trees in a path between the summer sun and the house. This provides shading for the house and glazing. When the autumn winds blow, the leaves fall from the trees, allowing the welcome winter sun indoors.
- Plant pine, cedar, or other evergreen trees to the north and east to provide a windbreak. In treeless areas, wind and storm blocks can be made using rock outcroppings, hills, or even your neighbor's house.

These orientation strategies apply whether you are in the country or the city. If you are designing a home in a city, try to evaluate the sun's path over your lot and take into account any shading effects caused by future buildings. The Solar Pathfinder shown in Figure 8-10 works wonders if you are concerned about your home's view of the sun.

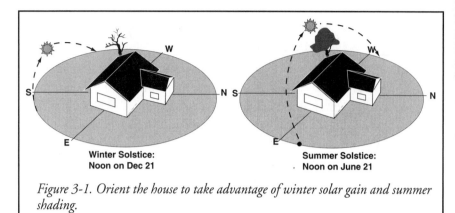

Winter Solstice: Summer Solstice:
Noon on Dec 21   Noon on June 21

Figure 3-1. Orient the house to take advantage of winter solar gain and summer shading.

Step 2 – Insulate, Insulate, Insulate

If you haven't taken the time to review Chapter 2, "Energy Efficiency", go back and read it. A house that is underinsulated and prone to air leakage will quickly put an end to your quest for energy efficiency. **Remember that it is far less expensive to conserve energy than to acquire more energy to heat a substandard house, even if the energy is free!**

If you live in an area where summer air conditioning loads are high, consider adding radiant barrier insulation in the attic, as discussed in Chapter 2.

Step 3 – Window Design

Windows are the passive solar collectors of the house. Vertically oriented glass will catch the low winter sun and miss the high summer sun angle. This means you can create a "normal" looking home and still get maximum solar efficiency.

- Place the majority of the desired glazing on the south side (long axis) of the house.
- Choose windows that are of the highest quality you can afford. Refer to Chapter 2 for information on window construction. Remember that triple-glazed glass with Krypton gas fill is the most efficient available, reducing nighttime energy loss.
- Do not use low-E glass on the south side of the house or on any window that you wish to use as a solar collector.
- Ensure that window units that can be opened are installed on the side of the house facing the prevailing winds as well as on the opposite side of the house. You can open these windows to provide nighttime cross-flow cooling of the home.

Figure 3-2. This home, located in Eastern Ontario, is designed to capture the maximum amount of winter sun energy, while avoiding the high-angle summer rays.

Figure 3-3. This lovely veranda also acts as the south window summer shading system. The roof overhang was designed to allow sunlight to enter the house structure from mid-September through mid-March.

Figure 3-4. This newly built "earthship" design is very heavily glazed by northeastern standards. It will require plenty of shading and cross-flow ventilation to keep it from overheating in summer. (Note the photovoltaic panels mounted to the right of the windows.)

- Limit the amount of glazing on the north, northeast, and northwest sides of the house.
- Do not overglaze the south side. Excess window area equates to excess heat.

Step 4 – The Finishing Details

These finishing touches will assist in making the job work "just right":

- Provide proper home ventilation using an air-to-air heat exchanger unit as outlined in Chapter 2.
- Provide sun shading over the east-, west-, and south-facing windows as required, to prevent the summer sun from heating the structure.
- Use summer sunscreen blinds similar to the units shown in Figure 2-10.
- Install a thermal absorber and buffer system in the living area. The sun will quickly overheat a properly insulated house, making it unbearable. A solar mass such as cement/tile flooring or walls, finished in dark colors, will absorb heat, helping to moderate temperatures. During the evening hours, this stored energy is radiated back into the room, helping to keep temperatures constant.

The key to ensuring that your passive solar design does not cause extreme cycling of interior temperatures is to balance the solar collector area (windows) with the thermal storage and buffering mass of the floor and wall systems. When these components are not in balance the results are very similar to a greenhouse: hot during the day and cold at night.

An educated architect or heating contractor will be able to assist you in developing a heat-loss calculation for your home. Alternatively, a home inspection auditor who is capable of performing an air leakage test may be able to assist with this calculation. Armed with this information, you will be able to calculate what percentage of the yearly heating load can be expected from solar energy. With the sun providing approximately 1000 W per square meter of window area, it's a shame not to take advantage of this free energy source.

3.2
Active Solar Air Heating

Figure 3-5. The Solarwall® system capitalizes on the fact that a building's exterior cladding will be exposed to the energy of the sun and might as well capture that energy for fresh air heating and ventilation. (Courtesy Conserval Engineering Ltd.)

A simple, unique solar heating concept is produced under the trade name Solarwall® by the firm Conserval Engineering Ltd. This cladding material is mounted on the south, east, or west walls. Tiny perforations allow outdoor air to travel through the exterior surface. During the day, as outside air passes through the panel and along its inside surface, it absorbs the sun's energy, warming the air which in turn rises. The preheated, warm air is then drawn into the building's ventilation system, reducing the load on the heating system.

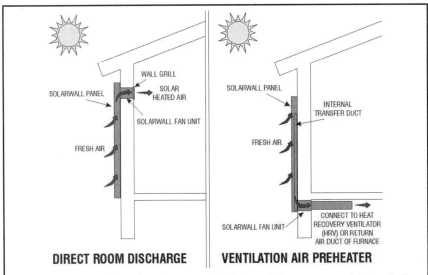

Figure 3-6. In residential applications, the Solarwall® system may be installed as a direct room discharge unit or as the preheater for the intake of a furnace or heat recovery ventilator system.

Figure 3-7. This view of a partially completed home shows the Solarwall® prior to the installation of the fieldstone siding desired by the homeowner. The design will work well with the stone facing, providing a very aesthetic finish.

During sunny days the Solarwall® preheats intake air by between 30°F and 50°F (17 and 28° C), depending on the desired flow rate. Snow can reflect up to 70% of solar radiation, into the Solarwall®, increasing the system's performance further. The system also works on cloudy days, as diffused light can provide up to 25% of the total radiation available on a sunny day.

The Solarwall® can even help cool your building in the summer. The cladding structure shades the interior wall and any hot air that accumulates at the top of the unit escapes through the perforations, keeping the wall surface cooler than it would be if it were exposed directly to sunlight. Electrically (or manually) controlled dampers allow unheated air into the building, maintaining air quality. The dampers can also switch to the wintertime position, providing preheated air intake.

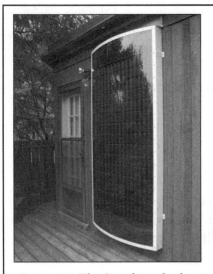

Figure 3-8. The Cansolair solar heating unit may be installed as a retrofit device on the south-facing wall or roof. Using a thermostatically controlled fan unit (shown above right), the system can provide preheated intake air to the building ventilation system or directly to a room The unit produces approximately 2,900 kWh of energy per year based on 2,200 hours of sunlight. At a hydro cost of 10 cents per kWh, this equates to almost $300 per year of free energy.

Figure 3-9. The Solarwall® system has been adapted to act as a metal roof structure as well as an air solar thermal collector. (Courtesy Solar Unlimited Inc., Utah)

A slightly modified version of the vertically installed Solarwall® is the roof-mounted system shown in Figure 3-9. If you are building a new home or planning to re-roof an existing house, the incremental capital cost may be negligible.

An Alternative to the Solarwall® system is a retrofit product known as the Cansolair unit. The Cansolair is a cross between the Solarwall® system and a greenhouse, contained within a package of the same size as a sheet of plywood. The heated air may be delivered to the home's heat recovery ventilator or via a thermostatically-controlled room ventilation fan shown in Figure 3-8.

The economic case for solar thermal heating systems depends on many factors including the age of your home, its heating demand, and its thermal efficiency. There is little point in installing a system that may only provide a small percent of total heat demand because of a thermally leaky house. On the other hand, an energy-efficient home that has a correspondingly lower heat demand may receive a significant portion of its total yearly thermal energy from such a system. It will be necessary to compare the heat loss of your home with the heat production of the solar thermal unit to determine how much purchased energy, in the form of oil or gas for example, the system will offset. This is a complex calculation and is best left to professional heating and cooling contractors and their computers.

3.3
Solar Water Heating

Figure 3-10. Solar thermal water heating systems are used throughout the world and with good reason. Approximately one-third of the energy bill for a home is related to hot water heating costs. Almost everyone is aware that Arizona and Florida can achieve huge savings with solar systems, but what about the majority of people who live in more northerly climates? It may come as a surprise, but if you live in Anchorage, Alaska or Inuvik in Canada's arctic you can still achieve a 35% reduction in hot water heating costs by using a solar thermal water heating system. (Courtesy EnerWorks Inc.)

When you ask most people about solar systems almost everyone will talk about heating for domestic hot water supply, and with good reason. Approximately-one third of the energy bill for a home is related to hot water heating costs. According to a Florida electrical utility, for every $1000 you pay in energy costs, $300 goes towards water heating. This figure is based on 2002 data as shown in the pie chart in Figure 3-11.

While everyone expects solar energy to make economic sense in the tropical or desert areas of the southern United States, what about the populated areas to the north, including Canada and its arctic regions? The maps shown in Figure 3-12 indicate the percentage of hot water heating energy that can be achieved in different geographic locations. Throughout the densely populated northeastern section of the United States and Canada, homeowners can achieve an average of 55% of their water heating demand from solar. If you happen to live in Anchorage, Alaska or even Inuvik in Canada's arctic region, you can still achieve an impressive 35% average savings in annual water heating costs.

Figure 3-12. These maps of Canada and the United States show the average annual solar energy that can be captured to reduce home hot water heating costs. Solar thermal hot water systems are one of the most economic means of putting the sun's energy to work. A typical family of four can expect to receive a return on investment of approximately 10% to 14% if they are located in an area with a minimum of 50% solar heating penetration. For those lucky enough to live in the Arizona desert or sunny Florida, the return on investment can exceed 20% according to industry figures. (Courtesy EnerWorks Inc.)

You may be wondering if your house will have hot water only for the percentage of time indicated in the maps in Figure 3-12. All solar thermal systems work in conjunction with your existing hot water heating system, providing "pre-heating" of the water supply. For example, if the desired hot water temperature is 140°F (60°C), the solar water heating system preheats the incoming cold water supply to 110°F (43°C), reducing the fuel demand of your hot water heater, while maintaining adequate supplies of hot water.

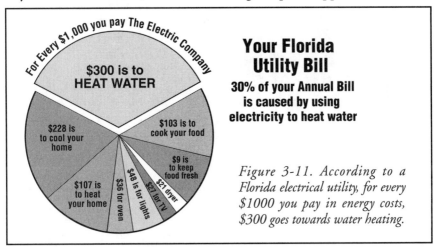

Your Florida Utility Bill

30% of your Annual Bill is caused by using electricity to heat water

Figure 3-11. According to a Florida electrical utility, for every $1000 you pay in energy costs, $300 goes towards water heating.

Before Installing a Solar Thermal System...

Most people practice their Luciano Pavarotti while taking a shower; I expect that the engineering minds at Renewability Energy Inc. focus on just how much hot water is actually going down the drain. In fact the hot water used for household tasks such as laundry, dishes, showers, and baths requires enormous amounts of energy, but much of the heat contained in the water is not extracted before the job is completed. If the drain water were cold, the energy efficiency of the task that uses the hot water would be 100%. However, this is never the case, and a considerable amount of energy actually goes down the drain in the form of warm waste water.

The Power Pipe™ heat recovery system is an amazingly obvious device which transfers energy from warm waste water to incoming potable cold water, which in turn supplies the home's hot water heater. As shown in Figure 3-13, a copper drainpipe is mounted in a vertical orientation, allowing waste water to pass through on its way to the sewer. The vertical orientation is important in order to ensure that drain water "skins" along the inside diameter of the pipe as it makes its way to the sewer, increasing the surface temperature of the pipe. A second smaller pipe is wrapped in a tight spiral around the vertical drainpipe. The bottom or inlet side of the spiral pipe is connected to the incoming cold water supply. As the warm drain water pours down the larger vertical pipe, heat is transferred to the spiral water supply pipe. Potable water exiting from the top of the spiral assembly is thus pre-warmed and may be sent to the hot water heater or plumbing fixtures, reducing their energy requirements (see Figure 3-14).

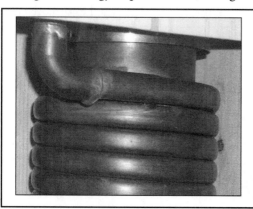

Figure 3-13. A close-up view of the top section of the drain and spiral wound supply pipes of the Power Pipe system. After installation the entire assembly is wrapped in insulation, preventing heat loss and increasing thermal efficiency. (Courtesy Renewability Energy Inc.)

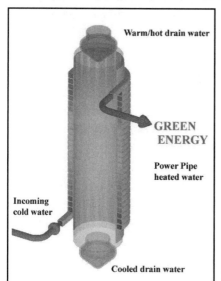

Figure 3-14. As warm waste water pours down the large vertical pipe, heat is transferred to the spiral wound water supply pipe which in turn feeds a hot water heater or the cold side of plumbing fixtures. (Courtesy Renewability Energy Inc.)

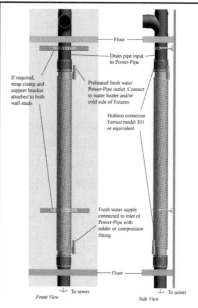

Figure 3-15. Although it is possible to retrofit your home with a Power Pipe heat recovery system, it is far easier to do the installation during construction. (Courtesy Renewability Energy Inc.)

Understanding the Basics of Solar Domestic Hot Water Systems

Solar hot water heaters work in conjunction with your existing fossil-fuel or electrically operated water heating tank, supplying a percentage of your total hot water heating requirements. They will reduce your hot water heating costs, greenhouse gas emissions, and smog and atmospheric pollutants, contributing to a cleaner and more sustainable environment.

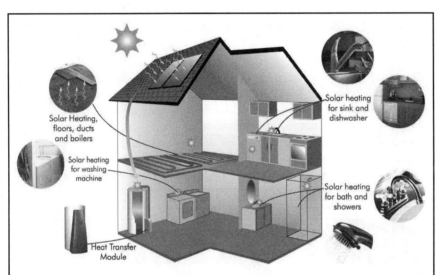

Figure 3-16 is a simplified view of a typical SDHW system. The typical urban solar water heater consists of roof-mounted panels which capture the sun's energy as heat. A circulator pump causes a transfer fluid to absorb this energy and pass it to a heat exchanger which in turn heats water in a storage tank. Solar-heated water is most often used in the kitchen, laundry, or bathroom, although it can also provide heat for space, pool, and spa applications as discussed later in this chapter. (Courtesy EnerWorks Inc.)

Energy Collection

Solar collectors are best mounted on the roof, facing solar south and mounted at an angle approximately equal to your geographic latitude in much the same manner as electricity-producing photovoltaic panels (see Chapter 8). Solar collectors are available in a number of different configurations, although flat plate designs (see Figures 3-10 and 3-17) are the most common. Evacuated tube collectors (see Figures 3-18 and 3-19) make up the balance, although newer collector configurations show up on the market from time to time (see Figure 3-20).

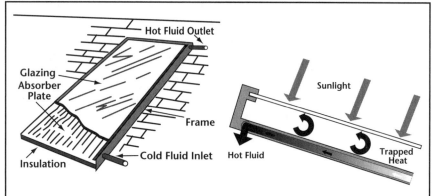

Figure 3-17. Flat plate solar collectors are miniature greenhouses which capture sunlight, trapping heat between a glazing material and an insulated rectangular box. A series of thin pipes attached to a dark-colored heat absorbing plate is placed inside the collector box. A circulating fluid such as water or antifreeze passes through the pipe assembly and heat collector plate, absorbing the trapped heat. If the heated fluid is potable water, it is supplied directly to the water heating unit. In severely cold climates an antifreeze solution is used instead of water, requiring the use of a heat exchanger to transfer heat energy to the hot water system.

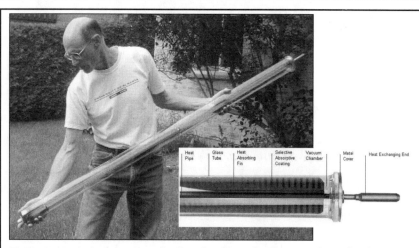

Figure 3-18. A single evacuated tube solar thermal collector. A heat absorber plate is bonded to an apparatus known as a heat pipe. The assembly is placed inside a glass tube which is sealed and evacuated, creating a vacuum. The collector plate absorbs the sun's energy, causing a fluid inside the heat pipe to boil. Heat is then transferred to a condenser bulb mounted at the top of the unit. If you have ever made coffee and placed it inside a thermos bottle you know how long the drink stays hot. In a similar manner, heat cannot escape from the evacuated tube collector, providing a system that is not affected by extremely cold weather or high water temperatures. (Courtesy Carearth Inc.)

Figure 3-19. Evacuated tube solar collectors are most commonly mounted to a "header" in groups of ten. The transfer fluid flows through the header, absorbing heat from the condenser bulb located at the top of the evacuated tube. The header assembly is well insulated, preventing heat loss.

Figure 3-20. Solar thermal collection systems can also go high-tech. This Power-Spar™ from Menova Engineering Inc. uses a spherical mirror to concentrate the sun's energy directly onto a fluid transfer header, which has the effect of amplifying the sun's energy. Additionally, the Power-Spar™ contains a unique solar tracking unit which ensures that the mirror is correctly aimed at all times. The Power-Spar™ may also be fitted with photovoltaic panels to produce electricity at the same time as heat energy.

Flat plate solar collectors tend to be less expensive than their evacuated tube cousins, explaining their relative popularity. However, for the same energy output flat plate collectors tend to require more surface area for mounting than evacuated tubes. In addition, evacuated tube configurations operate at higher thermal efficiency and are therefore better suited for operation in extremely cold environments. If neither of these considerations affects you, the selection of which collector to use can be based strictly on price.

Regardless of which solar thermal collector you choose, it must have an unobstructed view of the sun between the hours of 9 a.m. and 3 p.m. throughout the year. Before purchasing a system, ensure that your site provides the necessary solar exposure. In addition, the collectors must be mounted so that they are facing as close to solar south as possible, bearing in mind that a phenomenon known as magnetic declination must be considered when using a compass for alignment. Magnetic declination maps provide a correction to compass readings in order to locate true north and solar south. Consult Appendix 4 to determine the magnetic error in your location.

Remainder of System

There are numerous types of solar water heating systems available, including the batch storage units shown in Figure 3-21. Although common throughout the developing world, these units are not considered aesthetically pleas-

Figure 3-21. Batch solar and passive open-loop systems use natural convection to circulate water through the collectors and water storage tank. Although not common in North America, these units are ideally suited for three-season cottages or where freezing temperatures and visual aesthetics are not a concern. (Courtesy Carearth Inc.)

ing to the North American eye, and they are susceptible to freezing. They are, however, extremely simple and relatively inexpensive and may be ideally suited for three-season cottages in northern environments or full-time residences in the south.

The most common configurations in use today are active closed-loop systems (Figure 3-23) that use a pump, temperature sensors, and electronic controls to regulate fluid circulation through the solar collector(s) and heat exchange mechanism.

The drain back system shown in Figure 3-22 features a standard hot water tank with a non-pressurized secondary water storage tank mounted on top. Thermal sensors mounted at the solar collector panel and the hot water tank are connected to an electronic device known as a temperature differential controller. This unit measures the difference in temperature between the solar collector and the hot water tank. When the collector temperature is higher than the hot water tank *and* below the desired water temperature of the system, the circulator pump is activated, transferring heat to the tank. When the solar collector cools, for example at night or on a cloudy day, the differential controller will turn off the circulation pump, causing water to flow back into the secondary storage tank. As the main water heater tank is equipped with a heat exchanger, water in the secondary storage tank does not come into contact with the potable hot water supply.

Drain back systems are relatively simple and theoretically immune to freezing. However, under unusual conditions

Figure 3-22. The drain back system is relatively simple and theoretically immune to freezing, making it a popular choice for domestic solar thermal hot water systems in moderate climates. Water contained in a storage tank is circulated through the solar collector and the hot water tank, which is equipped with a heat exchanger unit. When an electronic controller stops the circulation pump, water will drain back into the storage tank, preventing freezing. (Courtesy Alternate Energy Technologies, LLC)

it is possible that water could freeze within the solar collectors or the plumbing before draining back into the secondary storage tank. For extremely cold climates, the closed-loop glycol system is preferred.

The closed-loop glycol system is the preferred configuration in areas where freezing is a concern. A mixture of food-grade propylene glycol antifreeze (it must be safe; soft ice cream contains 30% of the stuff) and water continuously remains in the solar collectors, plumbing lines, and heat exchanger. A differential electronic controller and temperature sensing devices are connected to an alternating current circulation pump in the same manner as in the drain back system described above. It is also possible to replace

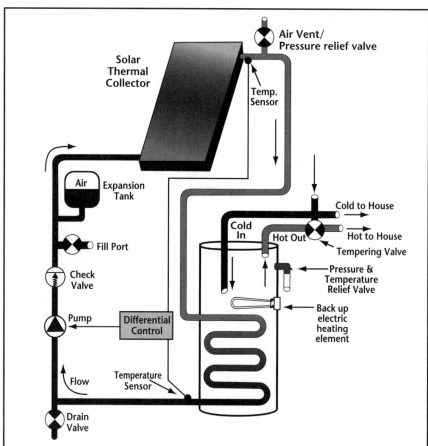

Figure 3-23. The active, closed-loop system shown above uses a heat-collecting antifreeze fluid which is circulated through the solar collectors and heat exchanger. Energy is stored in either a single-storage water heater tank (indicated) or in a two-tank configuration, according to the manufacturer's design.

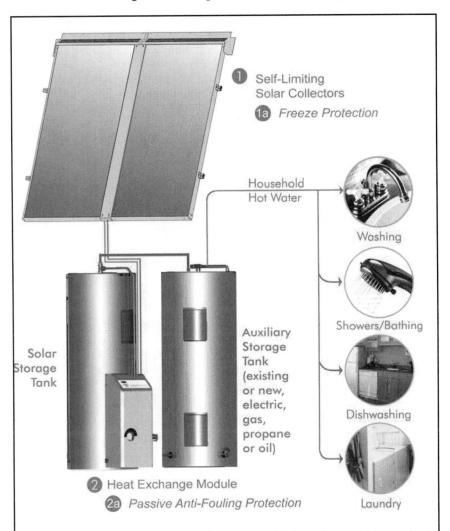

1 Self-Limiting
Solar Collectors

1a *Freeze Protection*

Household
Hot Water

Washing

Showers/Bathing

Solar
Storage
Tank

Auxiliary
Storage
Tank
(existing
or new,
electric,
gas,
propane
or oil)

Dishwashing

2 Heat Exchange Module

2a *Passive Anti-Fouling Protection*

Laundry

Figure 3-24. A mixture of food-grade propylene glycol antifreeze which remains in the solar collectors and plumbing lines at all times makes the closed-loop glycol system completely freeze-resistant in any climate. A differential electronic controller and temperature sensing devices activate a circulation pump, transferring energy to an existing hot water tank via a heat exchange module. (Courtesy EnerWorks Inc.)

the alternating current circulation pump with a direct current model connected to a small photovoltaic panel mounted in the vicinity of the solar thermal collectors. This configuration eliminates the need for temperature sensors and the differential electronic controller by activating the circulation pump whenever the sun shines.

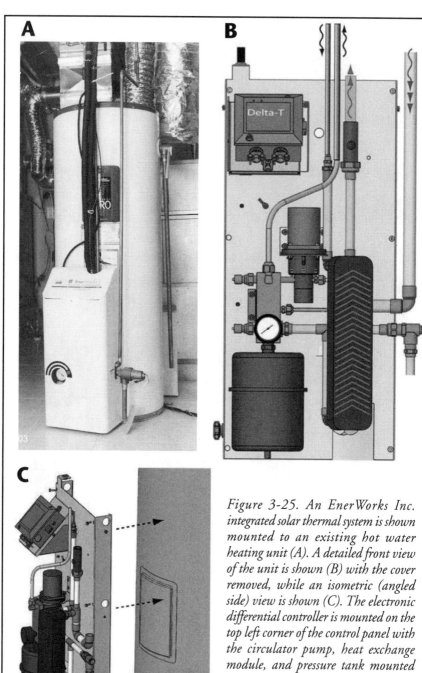

A

B

Delta-T

C

Figure 3-25. An EnerWorks Inc. integrated solar thermal system is shown mounted to an existing hot water heating unit (A). A detailed front view of the unit is shown (B) with the cover removed, while an isometric (angled side) view is shown (C). The electronic differential controller is mounted on the top left corner of the control panel with the circulator pump, heat exchange module, and pressure tank mounted below. (Courtesy EnerWorks Inc.)

Figure 3-26. The Grundfos brand of centrifugal-type hydronic heating pump is commonly used in solar hot water systems because of its low power consumption, high reliability, and low maintenance. Other manufacturers such as Taco and Hartell provide similar models.

Circulation Pump

Centrifugal-type hydronic heating pumps are used in solar hot water heating systems to circulate a heat collecting fluid. They are chosen because of their low power consumption, high reliability, and low maintenance. For long life it is recommended that bronze or stainless steel pumps be used in this application even though they are slightly more expensive than cast-iron models.

These pumps are readily available from any plumbing supply store that specializes in hydronic heating systems.

Plumbing

Solar thermal systems are generally plumbed using standard domestic copper plumbing pipe which is wrapped in an insulation sleeve material such as Rubatex brand Insultube. An alternative material is PEX tubing manufactured for the hydronic (in-floor) heating industry.

To minimize heat loss it is recommended that plumbing lines be routed inside the building envelope as soon as possible after leaving the solar thermal collectors. If you are planning on building a house, it is a good idea to pre-plumb lines from the location of prospective solar collectors to your mechanical room. The cost is insignificant and will save a considerable amount of grief during a future installation process.

Check Valve

A one-way check valve is installed on the outlet port of the circulation pump to prevent the convection flow of warm fluid from the hot water storage tank to the cold solar collectors at night. Check valves may be spring or gravity operated. Gravity-operated check valves must be installed in the orientation shown in the installation manual. In applications using an AC circulator pump, the spring-operated check valve is preferred, as it is less susceptible to internal leakage caused by minute particles circulating in the heat collecting fluid.

Fill and Drain Ports

One fill and one drain valve must be provided to allow the filling of the circulation system with water or glycol solution. In order to fill and pressurize the circulation system, a positive-displacement high-pressure pump is required to overcome the vertical "head pressure" related to the force of gravity and the expansion tank air pressure.

Figure 3-27. An expansion tank allows the water or glycol solution in the closed-loop system to expand and contract as a function of its temperature.

Expansion Tank

An expansion tank allows the water or glycol solution in the closed-loop system to expand and contract as a function of its temperature. Without an expansion tank the buildup of pressure would cause the plumbing lines to burst when the circulation fluid is heated by the sun.

Hydronic expansion tanks such as the model shown in Figure 3-27 contain a rubber bladder which is in turn connected to the plumbing circulation

Figure 3-28. A pressure relief valve will prevent an explosion of the plumbing lines in the event of an excessively high pressure buildup.

Figure 3-29. An air vent is installed at high points in the circulation system and is used to remove trapped air (left). A pressure gauge (right) indicates that the circulation system is operating within acceptable pressure limits.

lines receiving the circulation fluid. With the tank empty the air chamber is pre-charged to approximately 15 psi (103 kPa) using a bicycle pump or compressor. As the circulation fluid expands and contracts as a result of the changing temperature, the pressurized air bladder maintains a relatively constant pressure in the plumbing system.

Your hydronic heating component supplier will be able to determine the optimum size of expansion tank based on the quantity of circulation fluid in the system.

Pressure Relief Valve

Pressure relief valves are designed to prevent the catastrophic explosion of plumbing lines in the event of excessively high pressure buildup. A setting of 50 psi (345 kPa) is common for most hydronic heating systems. All pressure relief valves including the model shown in Figure 3-28 are equipped with a drain port which must be connected to a floor or sanitary drain.

Pressure Gauge

A pressure gauge indicates that the circulation system is operating within acceptable pressure limits. The normal operating pressure is approximately 15 psi (103 kPa) and will vary somewhat depending on the temperature of the circulation fluid: the warmer the temperature the higher the pressure reading; the cooler the fluid the lower the pressure.

The pressure gauge may also be used to detect leaks in the circulation system. A slow continuous drop in pressure over time indicates a fluid leak.

Air Relief Valve

Air trapped in the fluid circulation lines will lower the heat exchange efficiency or stop the circulation of fluid altogether. To remove trapped air, a relief valve (Figure 3-29, left) is installed at high points in the fluid circulation lines and at the solar collector panels. During commissioning, these valves are opened, allowing the trapped air to escape to the atmosphere.

Atmospheric gases are often dissolved in the circulation fluid and make their presence known by a gurgling sound in the plumbing lines. If the circulation system is completely free of air there should be no sound at all. During the yearly maintenance review of the system it may be necessary to activate the relief valves to eliminate any trapped air.

Figure 3-30. The heat exchanger transfers energy from the solar collectors to domestic potable water in the storage tank.

Heat Exchanger

The heat exchanger transfers energy from the solar collectors to domestic water in the storage tank. Heat exchangers may be external to the storage tank, such as those shown in Figure 3-25, or internal, as shown in Figure 3-30.

Heat exchangers are available as single- or double-wall designs, indicating the number of mechanical barriers between the heat-absorbing circulation fluid and the potable water in the storage tank. All plumbing codes require the use of double-wall heat exchangers to prevent the contamination of potable water in the event of a leak.

Heat exchangers work when the circulation fluid is at a temperature higher than that of the water contained in the storage tank. Heat transfer is also affected by the surface area (size), thermal conductivity, and flow rating of the heat exchanger. If you are designing your own solar thermal system, contact a hydronic heating contractor to determine the appropriate size for your installation.

Figure 3-31. Temperature sensors are located on the hot fluid outlet line of the solar collector and the bottom or cold line of the water storage tank. They feed temperature information to the electronic controller module.

Antifreeze

To prevent the collector circulation fluid from freezing, an antifreeze solution of 50% propylene glycol mixed with 50% demineralized or distilled water is used. Common automotive antifreeze containing ethylene glycol must not be used because of its toxic nature.

The collector loop is filled with the antifreeze mixture using a positive displacement pressure pump. Sufficient fluid is introduced into the collector circulation loop to develop approximately 15 psi (103 kPa) of working pressure as indicated on the pressure gage.

Temperature Sensors

Temperature sensors are located on the hot fluid outlet of the solar collector and the bottom or cold heat transfer pipe of the water storage tank. The sensors, known as "thermistors," change their electrical resistance as a function of temperature, thereby feeding information to the electronic controller module.

The temperature sensor shown in Figure 3-31 is applied to the cold side of the thermal storage tank plumbing as indicated in Figure 3-23. Prior to attaching the temperature sensor to the plumbing lines, thermally conductive grease supplied with the controller module is added to ensure a proper temperature reading. The sensors are secured in place with a screw-type hose clamp as indicated in the picture.

Electronic Control Module

The brain of the solar thermal system is a device known as a differential temperature controller. This unit monitors the circulating fluid temperature at the outlet of the solar thermal collector and on the cold side of the water storage tank. When the temperature of the solar thermal collector outlet exceeds the cold-side temperature of the storage tank by a predetermined number of degrees, the circu-

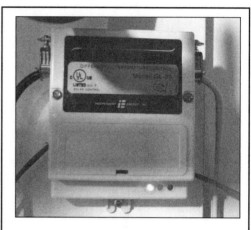

Figure 3-32. The brain of a solar thermal system is a device known as a differential temperature controller.

lation pump is started and the heat transfer process begins.

A differential temperature in the order of 10°F to 25°F (5°C to 14°C) is required in order for adequate heat transfer to begin. Electronic controls such as those shown in Figure 3-32 also contain a high-temperature safety limit or cutoff circuit which will shut the circulation system down once the water storage tank reaches a predetermined setting. The upper limit is adjustable and is normally set to 170°F (77°C).

Temperature Display

Two temperature gauges are recommended in order to measure the circulation fluid temperature at the inlet and the outlet of the heat exchanger, with a third gauge installed in the hot water storage tank as shown in Figure 3-33.

A temperature differential of approximately 20°F (11°C) across the heat exchanger indicates effective operation of the system. Monitoring the maximum

Figure 3-33. Two temperature gauges are recommended in order to measure the circulation fluid temperature at the inlet and the outlet of the heat exchanger, with a third gauge installed in the hot water storage tank as shown.

temperature of in the water storage tank provides an indication of the effectiveness of the high-limit cut-out circuit.

The House Side of the Plumbing System

So far we've concerned ourselves with the heat-absorbing side of the circulation system. The output of the solar thermal storage tank may be fed to a fossil-fuel or electrically powered water heater or directly to the house hot water supply.

Regardless of which installation is chosen, a temperature-limiting tempering valve should be added to the hot water outlet prior to feeding the household hot water circuits. The reason for this is that the solar thermal system will produce hot water of varying temperatures depending on the amount of sunlight. If the solar thermal system is incapable of meeting the desired temperature set point, a backup heating element installed in the solar tank or a regular secondary hot water heater will heat the water to the desired temperature. On the other hand, if there has been a significant amount of sunny weather the water temperature in the storage tank may exceed the desired set point temperature, scalding anyone who is not paying attention. A tempering valve will "mix in" an amount of cold water to ensure that the hot water supplied to the house is at the desired temperature, improving system safety.

Choosing and Installing a System

A qualified, experienced dealer will be able to help you select the water heating system that best meets your specific needs and can assist you in determining what state, municipal, or local utility tax incentives and rebates may apply. (You can also review the Database of State Incentives for Renewable Energy—DSIRE—at www.dsireusa.org.) If you have unpleasant recollections of some fly-by-night organizations selling solar thermal systems during the peak of the oil crisis, today's licensed professionals should be able to alleviate your concern.

It is wise to select dealers who are able to provide warranty service for both parts and labor and who have a satisfactory track record installing systems. Whenever possible, ask to see a previously installed system and try to discover if the owner is satisfied with both the installation and after-sales service.

Installing a solar thermal system is not the same as going to the hardware store to purchase a regular water heater. The dealer will have to perform a site evaluation, taking measurements and determining pipe routing and other details required to complete the installation. You will become an active part of this process.

Because you are making modifications to the structural and plumbing systems of your home, it will be necessary to install components that are certified by Underwriters Laboratories in the United States or the Canadian Standards Association to ensure fire and electrical safety. Purchasing approved products also lets you sleep at night knowing that product designs have met construction-quality standards. In addition, you must also obtain a building permit from your local inspection authority.

For local installation contractors and equipment suppliers in the U.S. contact the American Solar Energy Society at www.ases.org. In Canada contact the Canadian Solar Industries Association (CanSIA) at www.cansia.ca.

Operating a Solar System

Once the system is up and running you will be able to monitor its operation by observing the water temperature in the storage tank as well as the differential temperature at the inlet and outlet ports of the heat exchanger as discussed above.

You can also monitor the system for slow, minute leaks by watching the pressure gauge over a period of time. System pressure will rise and fall with a change in circulation fluid temperature; however the average "resting" pressure should remain constant over time.

In extremely sunny weather it is very likely that the system will produce more hot water that you can possibly use. When this situation occurs the electronic control module will stop the circulation pump, preventing the storage water tank temperature from rising above preset "over-temperature" limits. Circulation fluid trapped in the solar thermal collectors will continue to absorb energy, continuing to absorb heat.

This brings about a phenomenon known as "collector stagnation." If this condition persists the propylene glycol solution may prematurely break down, leaving a sticky, sludgy deposit in the plumbing components. To prevent this from happening manufacturers have developed a number of novel solutions. For example, EnerWorks has developed a "smart metal thermal actuator" which is installed on the underside of the company's flat plate solar collectors. During normal operation heat is drawn away from the collector by the circulating fluid and the actuator remains closed. When stagnation occurs, excess heat in the collector assembly causes actuators located at the bottom and top of the rear side of the panel to open, allowing fresh ambient

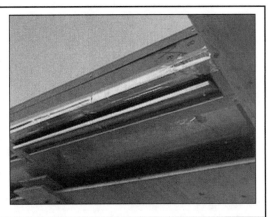

Figure 3-34. Manufacturers have developed numerous methods for preventing stagnation of the antifreeze circulation fluid. EnerWorks has developed a thermally activated vent which opens when stagnation conditions develop, admitting fresh ambient air. (Courtesy EnerWorks Inc.)

air to ventilate and cool the collector assembly. This operation is shown in the photograph and cross-section schematic in Figure 3-34.

An alternative to having this heat energy "wasted" when the water storage tank is filled is to include a bypass valve that will direct energy to other areas of the home that may be able to put it to work. For example, the hot fluid could be plumbed to a baseboard hot water radiator unit to provide auxiliary space heat in the home. Alternatively, a second heat exchanger could be plumbed into a hot tub or swimming pool, providing an excellent use for this energy. In either instance, your heating contractor will install the necessary valve and control equipment to ensure that domestic hot water is heated first with the auxiliary loads operated secondly.

An examination of the antifreeze circulation fluid is recommended at least every two years. There are two ways of testing this fluid. Samples may be sent for analysis to the Dow Chemical Company under its free annual fluid analysis program. The company will provide a complete summary report with a results table and a cover letter explaining the condition of the sample and outlining necessary maintenance procedures. (Contact Dow at: http://www.dow.com/heattrans/index.htm).

ACUSTRIP® Company, Inc. offers a simplified fluid test kit for propylene glycol antifreeze solutions. A sample of the antifreeze solution is placed in a clean glass jar. A test strip is immersed in the fluid and then compared to a color chart, indicating percentage propylene glycol concentration and pH (acidity) level. Ideal concentration levels are 50% propylene glycol to 50% demineralized or distilled water. The pH of the solution should be in the range of 8 to 10. If the glycol concentration is too low or the pH level too acidic, the system must be partially drained and new glycol added.

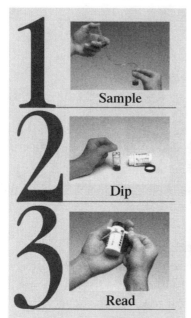

Sample

Dip

Read

Figure 3-35. ACUSTRIP® Company, Inc. offers a simplified fluid testing program for propylene glycol antifreeze solutions. This test kit should be used at least once every two years.

3.4
Solar Pool Heating

Heating a swimming pool using the heat of the sun is probably one of the most obvious and cost-effective applications for solar energy. According to the Canadian Solar Industries Association there are more than 250,000 swimming pools in Canada, of which approximately 40% are heated. It is further estimated that homeowners spend more money to heat their pools than their homes. Switching from fossil-fuel or electric heating to a solar thermal system would recoup the initial capital in less than 2.6 years, yielding a return on investment of approximately 38%. According to the GO Solar Company (www.solarexpert.com), using natural gas for pool heating costs in excess of $2000 for a typical swimming season even in sunny southern California.

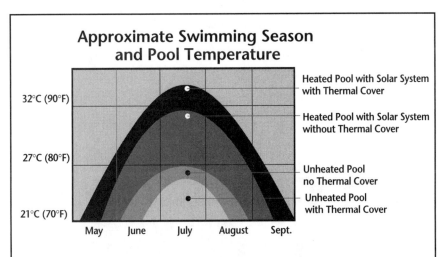

Figure 3-36. Using a solar thermal pool heating system and insulating solar blanket not only increases the pool temperature but can extend the swimming season by up to two months in the U.S. northeast and Canada.

The graph shown in Figure 3-36 indicates the increase in swimming pool water temperature as well as the extension of the swimming season when a solar thermal heater is used in conjunction with a solar blanket. A swimming pool that uses both of these items will be more comfortable and will save an enormous amount on heating bills. You will also eliminate tons of greenhouse gas emissions, literally! Figure 3-37 shows typical greenhouse gas emissions (in short tons: 1 t = 0.91 metric tonnes) during a single swimming season.

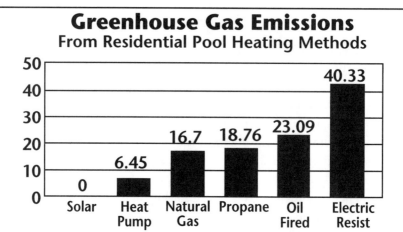

Figure 3-37. *Heating a swimming pool with fossil fuel is similar to throwing dollar bills into a lake: very expensive and not a really bright thing to do. Heating a swimming pool with electricity in Southern California creates 40 tons (36.3 tonnes) of greenhouse gas emissions in a single swimming season while costing the homeowner in excess of $2000. A solar thermal pool heating system can reduce both of these figures to near zero.*

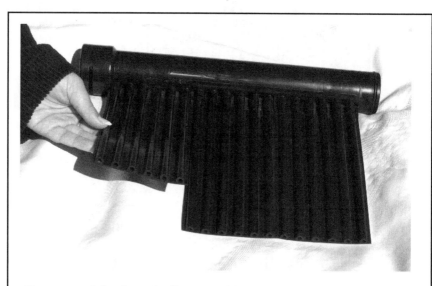

Figure 3-38. *Solar thermal collectors used for swimming pools may be as simple as these flexible rubber pad units manufactured by Enersol or the high-efficiency flat plate collector design discussed above. A typical collector surface area is approximately 60% to 70% of the swimming pool surface area.*

The Pool Heating System

To get the maximum benefit from your solar pool heating system, the collectors must be mounted on a south-facing roof or structure which is ideally angled to match your geographic latitude. It is also possible to place the collectors on the ground, although there will be a resulting drop in heating efficiency, requiring either the installation of additional collectors or an acceptance of reduced system performance.

The required collector area will depend on the surface area of the swimming pool, the desired water temperature, and the amount of shading from trees and neighboring buildings. As a general rule of thumb the collector surface area should be approximately 60% to 70% of the surface area of the pool. For example, a swimming pool that measures 10' x 20' (3 m x 6.1 m) will require a minimum solar collector area of:

Minimum solar collector area $= 10 \times 20$ *feet* $\times 0.6$
$$= 120 \, sq. \, ft. \, (11 \, m^2)$$

With few exceptions, the existing filter pump will be capable of circulating pool water through the solar panels. In areas where freezing is a concern, ensure that the system is equipped with drain spigots and properly sloped plumbing lines to ensure complete water drainage for winter storage.

Controlling the temperature of the swimming pool maybe accomplished in one of two ways. The first method incorporates nothing more complex

Figure 3-39. This three-season drain back solar pool heating unit extends the swimming season by two months and increases the water temperature when used in conjunction with a solar pool blanket.

than a timer to turn the circulation pump on and off in conjunction with sun up and sun down times. The advantage of this configuration is simplicity and low cost. The disadvantage is reduced pool circulation time and lack of control over pool water temperature.

An automatic solar pool heating control consists of an automatic diverter valve which channels water to the solar collectors based on differential temperature control and desired pool temperature. When the pool reaches the desired operating temperature, the diverter valve redirects the water flow from the solar collectors to the filtration system.

In either design an auxiliary fossil-fuel or electric pool heater may be added to the system. In this configuration the solar collectors will preheat the water before it reaches the backup heating unit. Fossil fuel or electricity will be used only to make up the difference between the solar thermal output temperature and the desired pool water temperature set point.

Figure 3-40. With few exceptions, the swimming pool pump should provide sufficient circulation of water through the solar collector panels. Temperature control may be as simple as a day/night timer turning off the pool water circulation at night.

3.5
Active Solar Space Heating

Before looking at how an active solar thermal system figures into the design, let's review the basic operation of a typical hydronic or hot water heating system.

Hydronic in-floor heating systems are becoming increasingly popular, in part because of the continuous, even warmth it provides throughout the home. As the heat is developed in the flooring material, even hard-to-heat products like ceramic tile stay nice and toasty for cold toes. Many people who suffer from airborne allergens find that their symptoms are greatly reduced with hydronic heating, as there is no dusty air being blown about the house through the centralized fan-forced heating system. Lacking a fan, hydronic systems are also very quiet and energy efficient.

The concept behind hydronic heating is quite simple. Heated water (or an antifreeze mixture) is pumped through a series of flexible plastic pipes located under the flooring or embedded in the insulated concrete slab of

Figure 3-41. With hydronic heating, hot water or antifreeze solution is pumped through plastic tubing located in or under your home flooring.

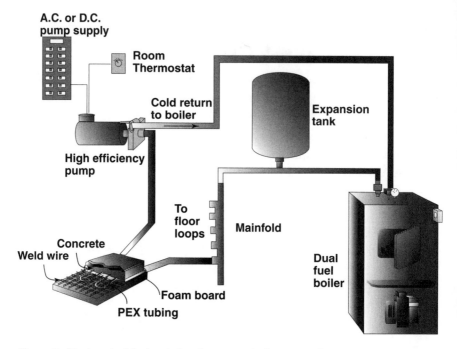

Figure 3-42. A typical hydronic heating system is shown in schematic view.

your house. A zone thermostat turns a very tiny, energy-efficient pump on and off as heat is required in each zone. For areas where freezing is a problem, environmentally friendly (propylene glycol) antifreeze is added to the water. This is also required for outdoor boiler heating systems described later in this chapter.

If you prefer to have a forced-air system or would like to add central air conditioning, a hot-water-to-air-heat exchanger can be used. This unit installs in the blower/ductwork assembly and transfers heat from the boiler to the circulating air in the same manner as a conventional forced-air furnace. A second air exchanger can be installed to add air conditioning to the mix.

Another advantage of hydronic heating concerns the boiler unit. Most boilers provide two "heating loops," one for hydronic space heating and the second for domestic hot water heating. These boilers can provide 100% of your space and water heating requirements using renewable resources such as biodiesel, wood, wood pellets, and even dried corn. Dual-fuel boilers are also available, typically combining an efficient wood-burning stove with an oil or electric backup energy supply.

Figure 3-42 illustrates a dual-fuel fired boiler which provides the energy for both space heating and, optionally, a second plumbing circuit for domestic hot water. The dual-fuel fired boiler has a wood-burning section on the top and an oil-fired burner on the bottom. Provided the wood fire is maintained, the boiler is programmed not to start the oil burner unit. This allows you to supply up to 100% of your heating needs from a renewable source.

Hot water or a water/antifreeze mix leaves the boiler and enters a manifold line which contains an expansion tank that is placed in the system to allow the water to expand or contract in volume, depending on temperature.

Figure 3-43. Active solar heating panels such as these evacuated tube thermal collectors can supplement your renewable energy hydronic heating system. The amount of solar energy developed during the winter and the heat loss of your home will determine the performance and cost effectiveness of the installation.

The manifold provides a number of outlets to feed the various zones or area heating loops, each equipped with its own circulation pump and thermostat.

The heating loops consist of, for example, a flexible plastic pipe known as PEX, manufactured by companies such as REHAU Inc. (www.rehauna.com). PEX pipe may be encased in insulated cement flooring over grade or affixed directly to the bottom of interior subfloors. Heating an insulated concrete slab provides a high level of thermal mass and helps to regulate the room temperature, thus providing much higher levels of comfort than central hot air furnace designs, where air temperature fluctuates as a result of the furnace cycling on and off.

PEX lines should be no longer than approximately 300 feet (91 meters), as the heat dissipates with distance. Cool fluid exiting the heating loop is drawn by a high-efficiency circulation pump and returned to the boiler water inlet. Hydronic heating contractors use circulation pumps from manufacturers such as Grundfos (www.grundfos.com) which are rated between approximately 40 and 80 watts at 120 volts

The controls for a hydronic heating system couldn't be easier. The pump is connected directly to a zone thermostat which in turn powers the pump on and off based on heat demand.

Heat loss calculations, PEX tubing layout, and other plumbing factors necessitate a discussion with your heating contractor to determine design and installation details.

Factoring in Solar Energy

As you learned above, a solar thermal hot water heating system such as the model outlined in Figure 3-23 will provide between 35% and 90% of your hot water requirements. What these figures do not tell us is when this energy is available and how it is dispersed over the year.

Appendix 5 and Appendix 6 illustrate the average sun hours per day for North America for the worst month and yearly average respectively. In the sun belt of the United States the average number of sun hours per day remains relatively constant throughout the year, meaning that solar energy is available throughout the summer and winter months. However, despite the availability of solar energy, home heating requirements in Arizona and Florida are minimal. In the northeastern United States and most of Canada solar energy is concentrated over the long summer hours. Once fall arrives and winter storms set in, sunlight hours per day drop dramatically, coincident with increasing heat demand.

A solar thermal system energy output is balanced to provide an *annual average reduction* in purchased hot water heating fuel as shown in Figure 3-12. Therefore, a solar thermal system will not meet 100% of the wintertime hot water requirement and will have little additional energy for space heating.

Many people believe this is a problem that can be corrected by size: simply add more panels and away you go. That's far too easy an answer. Firstly, a larger system would provide ample hot water for space and other domestic requirements at the expense of having an enormous amount of excess energy that could not be used in the summer. Secondly, the increase in capital cost for such a large configuration would almost certainly not be economical.

Perhaps storing some of this excess heat during the summer might solve the problem? Read on.

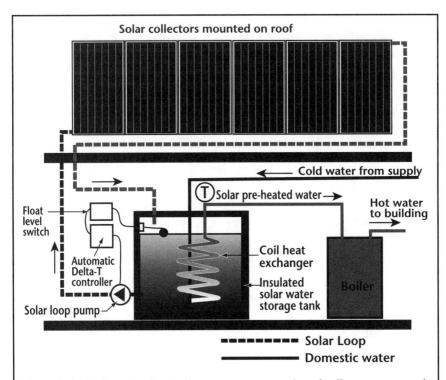

Figure 3-44. A large insulated solar water storage tank and collector array supply preheated water to a hydronic heating system. The practicality of such a system will depend to a large extent on the heat loss (energy requirement) of the building. It is not practical to store heat energy from one season to the next.

Thermal Storage and Heat Loss

It stands to reason that the amount of the energy required for space heating will be dependent upon how much heat the home loses. Heat loss is a function of how well the home is constructed and includes such variables as building envelope volume, insulation quality, and airtightness. In addition, the difference between indoor and outdoor temperatures adds a further complication.

There is also a difference between temperature and heat that must be understood prior to tackling the challenge of thermal energy storage. If you heat 1 cup (237 ml) of water over a flame, its temperature will rise more rapidly than 1 quart (1 L) heated by the same flame. Therefore, as the mass of water increases, so does the amount of heat energy required to achieve the same temperature.

In order to store heat, a mass (or weight if you prefer) of a substance will be required. Fortunately, water is an ideal substance as it has an unusual ability to store large amounts of heat energy, a phenomenon known as having a *high specific heat*. For example, the specific heat of lead is nearly 30 times lower than that of water. The measure of the specific heat of water is 1 BTU per pound per Fahrenheit degree (1 calorie per gram per Celsius degree). Expressed in English, this means that for every pound of water that is heated by 1°F, one BTU of energy will be added. Conversely, for every pound of water that is cooled by 1°F, one BTU of energy will be removed. (As the majority of heating contractors continue to work with British Thermal Units or BTUs, let's stick with this system. For discussion purposes, 1 BTU = approximately 1 kilojoule.)

For example, a tank of water weighing 1,000 pounds (120 gallons/ 455 L) which is heated to 50°F (28°C) above room or ambient temperature will have 50,000 BTUs of added energy. Assume for a moment that the tank is installed in a room which has a heat loss of 10,000 BTUs per hour. If we transfer 10,000 BTUs per hour from the storage tank to the room, it is possible to maintain a constant temperature for approximately five hours. Once all of the heat energy is removed from the tank, the room temperature begins to fall.

Stored Energy (BTUs) ÷ Heat Loss (BTU/Hr) = Available Heating Time (Hours)

and:

Heat Loss (BTU/Hr) x Desired Heating Time (Hours) = Heat Storage Required (BTUs)

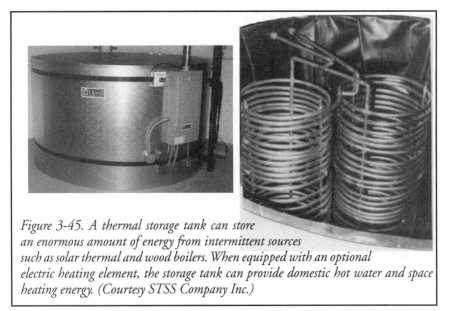

Figure 3-45. A thermal storage tank can store an enormous amount of energy from intermittent sources such as solar thermal and wood boilers. When equipped with an optional electric heating element, the storage tank can provide domestic hot water and space heating energy. (Courtesy STSS Company Inc.)

The simplest way to determine your home's heat loss is to have a heating contractor or home energy auditor run the calculations for you. It is not possible to generalize the heat loss for all homes and locations other than to say the more extreme the winter temperature and the larger and more poorly built the house, the greater the heat loss. A northern location provides fewer sun hours per day resulting in lower heat production and a colder climate requiring higher levels of heat energy.

A well-built home in wintry Montana may, for example, lose 400,000 BTUs of heat energy per day. In order to store this amount of energy for a one-month period the storage tank must be large enough to carry 12 million BTUs of energy:

400,000 BTU/day Heat Loss x 30 days' Desired Storage = 12,000,000 BTU

A storage tank containing water heated to 200°F (93°C) or 130°F (68°C) above ambient or room temperature would have to weigh approximately 92,000 pounds (42,000 kg) in order to store this much energy. To hold this mass of water the tank would have a capacity of 11,500 gallons (43,500 L). This is simply impractical.

On the other hand, if the heat loss is considerably lower, perhaps due to a more temperate climactic zone or better-built house, a smaller thermal storage unit may be practical. Furthermore, multiple energy sources can supply heat to one central storage tank. For example, the large storage tank shown in Figure 3-45 can be sized to hold up to 1,600 gallons (6000 L) of water with a weight of 13,000 pounds (5897 kg). Assuming a water temperature of 100°F above ambient, this equates to approximately 1.3 million BTUs of energy, enough to supply a family of four with a couple of weeks' worth of domestic hot water.

The manufacturer STSS Company Inc. of Mechanicsburg, Pennsylvania has developed a variety of sizes of these unique storage tanks, each with the ability to be shipped flat and then expanded at the final destination. Water/glycol solution can be circulated in a closed loop through the spiral heat exchanger, supplying domestic hot water and space heating.

An integral electric heating element and thermostat provide backup energy to maintain tank water temperature. Heat exchanger coils may also be routed to a wood- or oil-burning boiler, heat pump, and solar thermal collectors for additional energy input and storage.

The complexity of thermal storage units dictates that a comprehensive heat loss analysis of your home be completed prior to sizing the energy input sources and possible thermal storage system. This applies to solar thermal heating as well as to traditional heating sources such as fossil-fuel and electric heating supplies.

Given the difficulty in cross-season energy storage, solar thermal systems do not generally make *eco-nomic* sense in areas where winter heat demands are high.

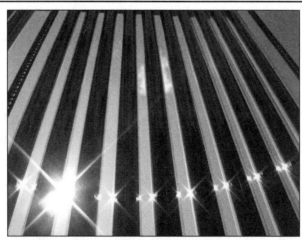

Figure 3-46. Solar thermal systems are the most economically feasible means of harnessing the sun's energy provided they are installed professionally and for the correct application. These evacuated tube solar thermal collectors provide the homeowner with free domestic hot water and a return on investment of 15% or more. (Courtesy Carearth Inc.)

3.6
Earth Energy with Geoexchange

Geoexchange technology is more commonly known by its many nicknames such as geothermal, heat pump, and ground-source heating. No matter what you call it, geoexchange technology offers a virtually endless supply of renewable solar energy that is stored just below the earth's surface since approximately 50% of the sun's energy is absorbed into the ground.

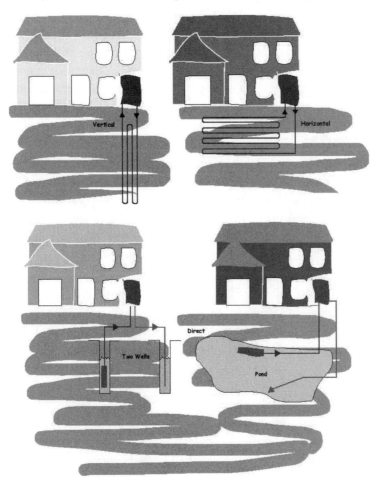

Figure 3-47. Geoexchange heat pumps are available in two broad categories known as open loop (bottom) and closed loop (top). Open loop systems extract heat directly from water pumped between two wells or from a lake or stream. Closed loop systems use a large array of pipes buried in the ground through which a circulating liquid returns heat to the compressor unit located inside the home.

The basic concept is easy to understand. In winter, heat energy is drawn from the earth through a series of pipes called a loop. An antifreeze solution circulates through the piping loop, carrying the earth's natural warmth to a compressor and heat exchanger inside the home, hereinafter called a "heat pump."

The heat pump concentrates the earth's energy and transfers it via duct work or a hydronic heating system for space heating requirements. If domestic hot water is also desired, the heat pump can supply energy directly to any thermal water storage tank similar to those in Figure 3-23.

Heat energy may also be extracted using open or closed loop groundwater transfer. In this configuration well or lake water is supplied directly to the heat pump and energy is extracted, with the cold discharge water returning to the lake or second well. Open loop systems may require annual coil flushing to remove mineral buildup if the source water is "hard."

In summer, the process is reversed; heat is extracted from the air inside the home and transferred back to the earth or water.

According to the U.S. Environmental Protection Agency and Natural Resources Canada, "**They are the most energy-efficient, environmentally clean, and cost-effective space conditioning systems available.**" Because geoexchange systems burn no fossil fuels onsite, there are no locally produced atmospheric emissions. If the electricity to run the geoexchange unit comes from hydroelectric sources, greenhouse gas and other emissions are reduced to zero.

There are approximately 500,000 installations in the United States and 35,000 in Canada resulting in annual energy savings of approximately 4 billion kWh of electricity, which eliminates 20 trillion fossil-fuel BTUs and slashing greenhouse gas emissions by approximately 3 million tons (2.7 million tonnes).

Air-to-Air Heat Pumps

An air-to-air heat pump operates in the same manner as a geoexchange heat pump with the exception that heating and cooling energy is transferred between indoor and outdoor air rather than between indoor air and the earth or well water. The common central air conditioning unit is an example of a unidirectional heat pump that can only provide space cooling.

Air-to-air heat pumps are not as efficient as their ground source cousins and are better suited to mild climates if large heating loads are anticipated. When operating in heating mode, energy extraction from the outside air diminishes as the temperature drops. At a temperature close to the freezing

Figure 3-48. The air-to-air heat pump is a cross between a central air conditioning unit and a geoexchange heat pump. Heating and cooling energy is transferred between indoor and outdoor air rather than between indoor air and the earth or well water. The outdoor component contains the compressor and heat exchange coil while the indoor unit contains the air handler, blower, and second heat exchange coil. Insulated refrigerant lines run between the two components.

point of water, the unit can no longer extract heat from the outside air. When this occurs, a standard resistance heating element, similar to those used in baseboard heaters and electric furnaces, is activated. Although heat provided by the electrical element is no more efficient than that of any other electric heating source, the overall seasonal heating efficiency of the heat pump *and* resistance heater must be taken into account, and it will always be greater than the heating element operating alone.

Measuring Heat Pump Efficiency and Value

Heat pumps use electrical power to operate a compressor which extracts approximately two-thirds of the home's required heating energy from the ground. Because this energy is free and the heat pump puts more energy into the home that it consumes from the electrical grid, heat pumps have an energy efficiency rating of greater than 100%. For every kWh of electricity supplied to the heat pump, between 2.8 and 6.7 kWh of energy are supplied to the home. This efficiency factor is known as the *coefficient of performance* (COP) and is calculated by dividing the heat output (in watts) by the energy input.

Heat pumps are able to operate in either a heating or cooling mode of operation. Therefore, when shopping for a system you must make coefficient of performance comparisons for both conditions. In the sun belt of the United States, cooling mode will be used much more frequently than heating mode. In the northeast and throughout much of Canada the opposite is true. When shopping for a heat pump, purchase the model with the highest overall energy efficiency, bearing in mind that is better to have higher efficiency in the most frequently used operating mode should price or model selection criteria permit.

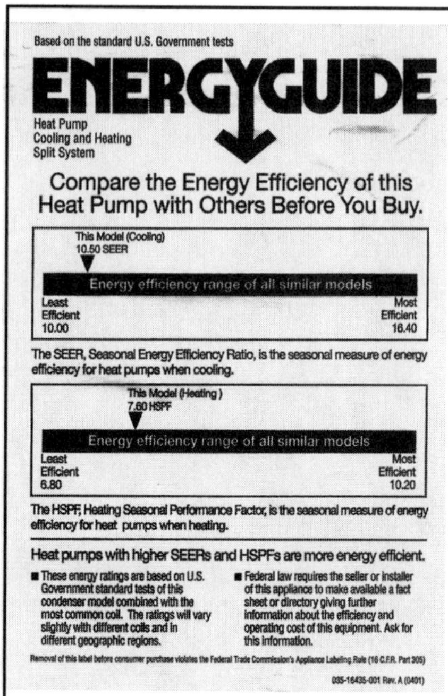

Earth energy systems have a reputation for being very expensive compared with traditional heating systems. However, this argument often lacks factual analysis. A conventional natural gas furnace and central air conditioning unit have approximately the same capital cost as a heat pump, but the cost of the gas lines and chimney must be factored into their installation cost.

The major difference in installation cost involves the ground loop or well water exchange unit as well as excavation or drilling labour. These costs can vary significantly depending on climactic conditions, heating and cooling demand, and which heat extraction loop technology is required. The larger the heating and cooling load of your home the larger the loop will be.

ENERGYGUIDE

Heat Pump
Cooling and Heating
Split System

Compare the Energy Efficiency of this Heat Pump with Others Before You Buy.

This Model (Cooling)
10.50 SEER

Energy efficiency range of all similar models

Least Efficient 10.00 — Most Efficient 16.40

The SEER, Seasonal Energy Efficiency Ratio, is the seasonal measure of energy efficiency for heat pumps when cooling.

This Model (Heating)
7.60 HSPF

Energy efficiency range of all similar models

Least Efficient 6.80 — Most Efficient 10.20

The HSPF, Heating Seasonal Performance Factor, is the seasonal measure of energy efficiency for heat pumps when heating.

Heat pumps with higher SEERs and HSPFs are more energy efficient.

■ These energy ratings are based on U.S. Government standard tests of this condenser model combined with the most common coil. The ratings will vary slightly with different coils and in different geographic regions.

■ Federal law requires the seller or installer of this appliance to make available a fact sheet or directory giving further information about the efficiency and operating cost of this equipment. Ask for this information.

Removal of this label before consumer purchase violates the Federal Trade Commission's Appliance Labeling Rule (16 C.F.R. Part 305)

035-16435-001 Rev. A (0401)

Figure 3-49. When shopping for a heat pump system, look for the EnergyGuide label in the United States and the EnerGuide label in Canada. Because these labels are developed by governments, other appliances list the most efficient first while heat pumps list the least efficient first! Heat pumps operate in either cooling or heating mode, necessitating an efficiency graph for each.

Moist, dense soil is better than light, dry, sandy soil for conducting heat into the system loop. Poor quality soil requires a correspondingly larger ground loop.

Several heat pump manufacturers provide software to aid in the assessment of soil type and loop size. In addition, a thorough understanding of local soil conditions is required.

After installation, heat pump operating and maintenance costs are considerably lower than those of conventionally fueled systems. Approximately two-thirds of the energy produced by a heat pump is free, reducing utility bills by 60% or more. Maintenance costs are lower because there is no combustion of fossil fuels to gum up the unit and all components are mounted indoors, protected from weather and vandalism. In fact, the only regular maintenance required is to clean the air filter, assuming you have a forced-air system.

According to Natural Resources Canada, independent research has shown that geoexchange heat pumps can be expected to perform for fifty years or more provided they are professionally installed.

It is impossible to calculate the capital and operating costs of a heat pump (or conventional heating/cooling system) without performing a heat

Figure 3-50. During site preparation a closed loop heat extraction field is excavated to a depth below frost line. (Courtesy Major Geothermal)

loss calculation for the home. As discussed in Chapter 2, an energy-efficient home will always be less expensive to operate, and upgrading an inefficient home will be more cost effective than installing a larger heating or cooling unit and having to pay excessive energy bills because of it.

Historically, the majority of general contractors and architects will choose the lowest first-cost mechanical heating/cooling system. The lowest-cost system will usually yield the least comfort, cost most to operate, and often be installed by the lowest-cost contractor. A mechanical contractor who is the least expensive installer usually can't afford to stay current with the proper training to understand how to design and install a well-balanced system and provide follow-up consultation and thorough warranty service. Given that a geoexchange heat pump will last for many decades, a quality installation is advised.

Figure 3-51. A heat extraction field prepared with a backhoe, requiring less expensive excavation and soil removal than the method shown in Figure 3-50. (Courtesy Major Geothermal)

Figure 3-52. This close-up view of the closed loop heat extraction pipes shows how they are anchored to the earthen walls prior to backfilling. (Courtesy Major Geothermal)

Figure 3-53. This picture shows a large heat exchanger being lowered into an artificial pond. Because source water does not flow through the heat exchanger piping, this design eliminates maintenance required because of mineral buildup and fouling. (Courtesy Major Geothermal)

3.7
Heating with Renewable Fuels

If you ask people to name all the renewable home-heating fuels, chances are they will stall immediately after they mention "firewood." In many parts of the developing world wood is as scarce as the proverbial hen's teeth, leaving people to search for alternatives such as dried peat or cow dung. Fortunately we don't have to chase Bossy with a shovel in order to lengthen our list of renewable heating options.

Chapter 5 examines ethanol and biodiesel as two renewable and clean-burning fuels for transportation requirements. For those heating with regular fossil furnace oil it may come as a surprise to learn that it is essentially the same thing as No.2 automotive diesel fuel or its renewable partner, biodiesel.

Biodiesel is a vegetable oil or animal fat derivative that is clean burning as well as carbon neutral, which means that it contributes no net carbon dioxide greenhouse gas emissions into the atmosphere. With North American production of biodiesel running in excess of 20 million gallons per year (76 million liters per year) and numerous states offering it as a clean alternative to fossil-fuel heating oil, demand is starting to heat up.

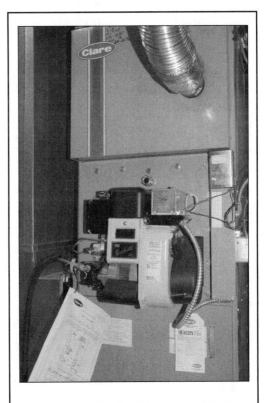

Figure 3-54. Biodiesel is a vegetable oil or animal fat derivative that is clean burning and carbon neutral, which means that it contributes no net carbon dioxide greenhouse gas emissions into the atmosphere.

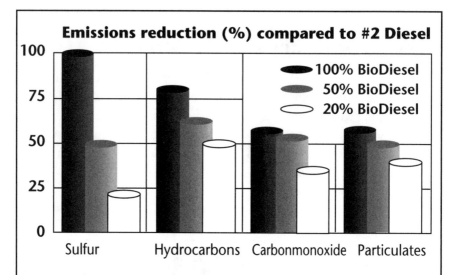

Figure 3-55. Biodiesel may be used in its pure form, designated B100 to indicate the percentage of biodiesel in the mix, or blended with fossil-fuel heating oil. This chart indicates the reduction in various atmospheric pollutants based on the blend concentration.

Wood, Wood Pellets, and Corn

Heating with wood goes back thousands of years. Europeans who settled in the New World along the east coast and in Pennsylvania experienced winters in North America that had a ferociousness never seen in Europe. Rapid population growth coupled with poor energy efficiency of both wood-burning appliances and homes caused heavy deforestation and left towns with a grimy, sooty pall. It's no wonder that the majority of the population eventually jumped on the oil, natural gas, and electric heat bandwagon. Write a check, set the thermostat, and you have instant comfortable central heating—what could be better? Why would anyone take a step "backwards" into the wood-heating scene?

Since cavemen started using fire, there have been several advances in wood-heating technology. These advances have included new wood-burning stove designs that increase the amount of heat recovered and at the same time lower environmental pollution. Besides saving the environment, these technologies can also save your back; when you couple the energy-conservation techniques discussed in Chapter 2 with clean, high-efficiency wood-burning appliances, you reduce the amount of wood you have to cut, split, pile, and burn.

Figure 3-56. Firewood stacked neatly in your garage or yard is like money in the bank. When properly burned in an EPA-certified wood heater it is a clean-burning energy-efficient heat source (Canada utilizes the EPA certification system). Wood pellets and dried corn also provide a clean-burning, renewable heat source with the added benefit of automatic stoking in many models of heating unit.

When you have decided on wood (which for the sake of brevity includes wood pellets and dried corn fuel sources) as your main or secondary source of home heating, check with your stove supplier or local building inspector for installation details. Since the installation of wood and pellet stove systems is complex and varies from region to region, no specific installation details are provided in *$mart Power*. It is highly recommended that a contactor or professional installer do the installation work. Use the details provided here to increase your knowledge and understanding of wood-burning appliances and to help you purchase the best unit to meet your needs.

Wood-Burning Options

Burning wood is not just about tossing a log into a fireplace or refurbished wood stove from the antique dealer and expecting to heat your house. It takes a high-tech wood-burning stove to heat cleanly and efficiently for a long period of time. There are several types of stove that fall into this category:

- Advanced combustion stoves
- Catalytic stoves
- Wood pellet and corn stoves
- Russian or masonry heaters
- Wood furnaces and boilers

Figure 3-57. This Vermont Castings (www.vermontcastings.com) catalytic wood stove easily heats this 3,300 square-foot, ultra-energy efficient home. Located in the very cold reaches of eastern Ontario, the home requires only two cords of firewood per heating season.

Designers and manufacturers of these units have as their primary aim the safe and clean extraction of every BTU of energy available from the wood you load into the stove. The reduction in frequency of wood loading lowers costs by reducing the amount of wood purchased or cut and processed.

The increase in efficiency comes first from creating an airtight burning chamber, causing controlled combustion, and then from developing a secondary combustion process, burning the smoke emitted from the fire before it reaches your chimney. (The masonry stove differs from this concept and will be discussed later in this chapter.) Although the EPA wood heater certification program was created to reduce air pollution, it resulted in added benefits like higher efficiency and increased safety. On average, EPA-certified stoves, fireplace inserts, and fireplaces are one-third more efficient than older conventional models. That means one-third less cost if you buy your wood and a lot less work if you process it yourself.

An airtight burning chamber is required to limit the amount of oxygen reaching the fire, lengthening the burning time and evening out the heat flow from the wood stove. This eliminates traditional fireplaces as efficient heat sources, as the energy they produce is negated by the massive amounts of cold makeup air, drawn from the outdoors, that is required to keep the

Figure 3-58. As this picture illustrates, there is an amazing array of wood-burning stoves and fireplaces available for purchase. A reputable dealer such as Embers located in Ontario, Canada is a good place to start your quest for home-heating energy self-sufficiency.

fire burning. A slow, smoky fire also increases airborne pollutants and the risk of chimney fires as a result of unburned fuel coating the chimney as creosote.

The idea of burning the smoke before it reaches the chimney may sound like snake oil, but it's true. Modern wood-burning appliance technology is aimed at developing ways of capturing the unburned fuel present in the emission of smoke. Smoke is the result of burning wood decomposing into "clouds" of combustible gases and tar. Applying additional oxygen and heat or special catalyzing materials causes a secondary "burn." This results in increased heat output and lower atmospheric emissions as well as a decrease in fuel dollars and backache.

Advanced Combustion Units

Advanced combustion wood stoves expand on the concept of a simple box stove that has been in use for over a hundred years. Simple stoves allow the heat and smoke of the fire to travel in a direct path up the chimney. The advanced combustion stove places an "air injection" tube into the smoke path. Secondary inlet air is drawn in through the tube, increasing the oxygen content of the smoke and causing it to burn. The smoke then travels through a labyrinth path, radiating heat before it exits via the chimney. This approach

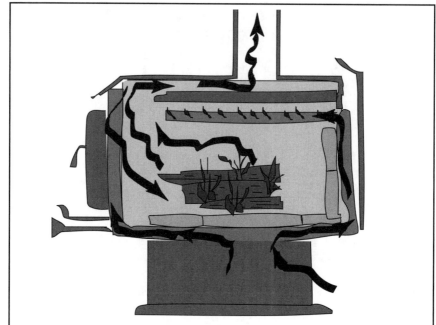

Figure 3-59. The advanced combustion stove relies on secondary oxygen intake and smoke burning to increase efficiency.

is like adding a turbocharger to a car's engine: free horsepower from otherwise wasted exhaust.

Catalytic Units

Once you understand how a catalytic wood stove operates, you can impress your friends by telling them how an automotive catalytic converter works. The process is really the same except that in the case of a car the unburned fuel is in the exhaust. Unburned wood smoke is routed through a catalytic device, a ceramic disk made with a myriad of honeycomb holes running through it. The disk is coated with a special blend of rare earth metals that have the unique feature of lowering the combustion temperature of the exhaust gases when mixed with a secondary supply of atmospheric oxygen.

As the smoke (or exhaust) passes through the catalytic device, it is mixed with oxygen from an air inlet and it ignites. If you watch the wood-burning stove's catalytic converter operating, you will see it glowing red hot and engulfed in flickering flames as the smoke is being consumed. A quick glance at the chimney on a crisp, cold January afternoon verifies the operation, with nothing but wisps of steam being emitted.

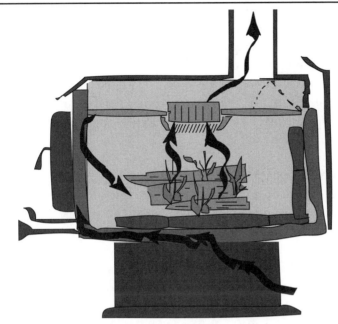

Figure 3-60. The catalytic wood stove uses a device made with special rare earth metals which allows wood smoke to burn in the presence of a secondary air supply.

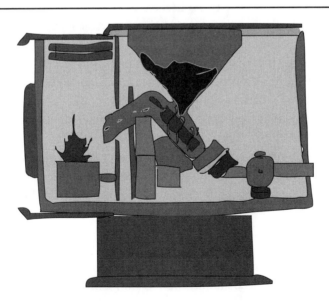

Figure 3-61. The wood pellet or corn stove uses a motor-driven auger to feed precise amounts of waste wood pellets into a fire pot.

The downside of catalytic technology is the requirement to move a control handle from the bypass to operating position once the stove has reached operating temperature. Although this may sound like a minor issue, laziness or neglect in operating this control will negate the energy-efficiency and emission-reduction capabilities of the stove.

Another consideration is the cost of replacing the catalytic converter after burning approximately 10 full cords (4' deep x 4' high x 8' long. / 1.2 m x 1.2 m x 2.4 m) of firewood.

Wood Pellet and Corn Stoves

Pellet stoves use dried corn and waste wood by-products from manufacturers of furniture, lumber, and other wood products. The waste material is ground and pressed together using naturally occurring resins and binders to hold the rabbit-food like pellets (Figure 3-56) together. As a waste product, biomass fuel pellets offer excellent synergy, heating your home while reducing landfill waste and needless greenhouse gas emissions at the same time. Pellets are convenient, as they are supplied in neat and compact dog food-size bags which can be stacked in your garage ready for use. Simply scoop a bunch into the hopper of the stove about once per day and the controlled feeding unit will automatically deliver the right amount of fuel to the burner. The only downside is: no electricity, no fire.

If you happen to have a grid-interactive renewable energy electrical system or backup generator (fueled with biodiesel of course), no problem. If not, most biomass stove manufacturers provide an automatic battery backup system to operate their units in the event of a power failure. This device operates in the same manner as an uninterruptible power supply (UPS) for a home computer. During normal operation the pellet stove operates from utility power and the UPS recharges simultaneously. When a power failure occurs, the UPS draws on a battery and inverter system to generate electricity, powering the stove. The UPS must be sized in accordance with the electrical load of the stove and the estimated number of hours before utility power returns.

Russian or Masonry Units

When we think of Russia, we think of winters in Siberia. In that neck of the world, serious heat is needed, and there is no fooling around with wimpy stoves. The masonry heater is a serious unit. It is designed with the firebox and chimney lined with refractory brick and the flue routed through a labyrinth path designed to slow the smoke on its way to the chimney. An external facing of brick, stone, or adobe completes the design and increases the mass of the unit.

When these units are operated, a fast, furious fire is ignited in the firebox. Smoke and its accompanying heat zigzag through the flue passages, giving up energy to the surrounding masonry work. Depending on the locale, one fast fire per day is all that is required to heat the masonry, with the stonework slowly radiating its stored heat into the house.

The masonry unit achieves its environmental passing grade by creating a hot, fast fire. Fast firing of a stove will generate the same amount of heat energy as a slow fire, but in a shorter amount of time. The resulting hot fire consumes more oxygen than a slow, smoldering fire, burning more completely and emitting fewer pollutants.

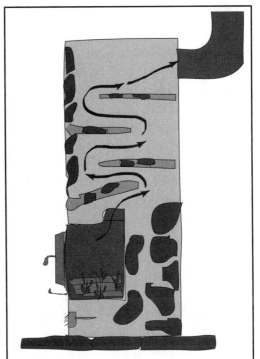

Figure 3-62. The size of a masonry stove makes it the centerpiece of the house. A popular design from the Old World, it provides heat long after the fire has burned down. (Courtesy Temp-Cast Enviroheat)

The downside of masonry heaters is their cost, the need to keep re-firing, and their size and weight. Assuming you have the space and the floor joists to support one of these big guys, they offer a warm and pleasing welcome to any home.

Wood Furnaces and Boilers

Wood furnaces and boilers come in two varieties: indoor and outdoor. Outdoor wood boilers are super-large wood stoves surrounded by a tank of water or antifreeze solution. Heat from the burning wood is transferred to the fluid, which is then pumped into the house for space and water heating. A control system located in the house regulates the temperature of the boiler by dampening the fire. Most units also provide an alarm signal to indicate when it's time to re-stock the unit. Furnaces, which provide warm air, are available as indoor units only.

Indoor wood boilers and furnaces are often dual-fuel fired, like the unit shown in Figure 3-64 from Benjamin Heating Products (www.benjaminheating.com). These units are very similar to outdoor models except for their size, emission of pollutants, and appetite for firewood. A major advantage of indoor, combination-fired boilers is the ability to supply heat even after the fire has burned down. The Benjamin unit contains both a high-efficiency wood-burning stove and an oil-fired backup burner which is only used when the wood-fired section cannot supply sufficient heat to maintain the desired boiler temperature. Units are also available with fan-forced heating which supplies a standard central air plenum or as hot-water boilers for use with hydronic heating.

Figure 3-63. Outdoor wood-fired boilers such as this model keep the wood and chips outside. Although these units are popular they are also inefficient, gobbling up vast quantities of wood while spewing out considerable amounts of atmospheric pollutants because of their inefficient burning chambers, which cause smoky, smoldering fires.

The Specifics of Wood Heating

The most popular style of heating system is the central warm air distribution design. A thermostat commands a centralized furnace to heat an air plenum or chamber. An electric fan circulates room air through the hot plenum to a series of ducts throughout the house. When the thermostat senses the room temperature is warm enough, the furnace turns

Figure 3-64. The combination boiler provides space and water heating in one compact unit. This model burns wood and oil efficiently and provides maximum flexibility of fueling options. (Courtesy Benjamin Heating Products)

Figure 3-65. The Multi-Heat automatic stoker boiler is especially suitable for effective and environmentally friendly firing with biofuels such as wood pellets and corn. The unit's efficiency of approximately 90% rivals that of high-quality oil and gas furnaces. The large hopper must be filled approximately once or twice per week depending on heat requirements. (Courtesy Tarm USA Inc.)

off, causing the house to cool, whereupon the cycle repeats.

Hot water or steam boiler systems replace the hot air with water. These systems use either radiator units to transfer heat to the room or hydronic, in-floor hot water pipes as shown in Figure 3-41.

A wood stove or masonry unit uses direct radiation and convection to distribute its heat.

When determining your choice of heating system, consider the following:

- Will the renewable heating system be your primary heat source or will it be supplementary? If it is to be your primary heat source, you must consider backup heating methods, particularly in colder areas subject to freezing.

- If you live in a more temperate area and travel little, and renewable heating is to be your primary heat source, consider pellet or masonry heaters. These models can provide sufficient heat for up to one-and-a-half days. The Tarm USA Inc. Multi-Heat stoker boiler shown in Figure 3-65 can operate for up to one week without refueling. Heating for longer periods of time will require a friendly neighbor to stock the stove or an alternate

Figure 3-66. A wood stove placed in a central location near open areas of the home facilitates airflow, helping to ensure an even flow of heat. Remember to plan access for wood supplies coming in and ashes going out.

heat source with automatic controls that activate when the fire dies.

Anyone who has ever been to a cottage that is heated with an old Franklin or cook stove cannot imagine using wood as a primary source of heat. These old stoves gobble up armfuls of wood and never seem to get the place really warm a lot of the time. At other times it's the opposite problem: there is too much heat and belching smoke, making you wonder if your cottage has been transported straight into the depths of hell.

Many people use wood or wood pellet stoves as their primary heating source. Combining one of these stoves with proper energy-conserving techniques and passive or active solar heating will give you the best heating synergies. The answer to avoiding the cottage nightmare is good planning and the use of quality equipment.

There are several systems for using wood as a primary heating source:

- Convection space heaters
- Wood-fired furnaces (single- or dual-fuel combinations)
- Hydronic or hot water in-floor heating

Figure 3-67. A wood pellet stove offers the look of a traditional wood-burning stove with the convenience of only having to add a few scoops of pellets into a hopper to provide equivalent heating capabilities. (Courtesy Harman Ltd.)

Convection Space Heating

Convection space heating is the most common type of wood heating installation. No electricity is required to operate this type of unit, which is doubly good when the grid fails. Locate the wood stove in a central part of the house, with open passages to the upstairs and other rooms. An example of a well-placed unit is shown in Figure 3-66, where heat from the stove can easily move around the home and up the open stairwell.

A common problem with space heating units occurs when people purchase a model too large for their needs. Make sure you discuss size with your dealer, and explain your requirements regarding primary or secondary heating as well as your floor plan. Most first-time wood stove users cannot believe that a small unit is capable of heating an entire house, but it is! On the other hand, if you are the sort who likes to wear shorts in January, maybe this isn't such a bad thing.

If your heating area is enclosed, or broken into many rooms, it is possible to use small "computer fan" units to blow air down hallways or between rooms. These units are available from wood stove dealers. Many building codes allow for grates to be cut between the ceiling of the heated room and the floor above, allowing natural air circulation.

Another option is to consider a pellet stove in tight locations. Because of their precise electronic controls, it is possible to lower burning rates below that of a typical wood space heater without degrading environmental operation and still provide a glowing hearth. An example of a Harman Ltd. space-heating pellet stove is shown in Figure 3-67.

A wood pellet stove hopper holds up to one-and-a-half days' worth of fuel. This offers the added advantage of not having to run back home to restock the stove or bugging your neighbor should you decide to stay out late at night.

Homes with an existing central furnace provide an automatic means of supplying backup heat. When the fire goes out the thermostat turns on and the furnace kicks in. This also provides an alternate means of moving the heat from around the wood stove. Figure 3-68 illustrates such a design. Heat from the wood stove pools at the ceiling level. An existing or additional cold air return or "suction duct" is located near the ceiling. The furnace fan is left in "manual" or low-speed mode, drawing excess heat from the room and circulating it throughout the house.

An effective alternative for homes without centralized air distribution or constructed with cathedral ceilings is to place ceiling fans in the heated area, blowing warm ceiling air downward and mixing it with cooler room air.

Figure 3-68. The air distribution provided by the central furnace fan provides the best method of moving heat away from the wood space heater. A ceiling fan is a good choice for cathedral ceilings or off-grid homes.

Wood-Fired Hot Air Furnaces

Wood-fired furnaces operate in the same manner as fossil-fuel units. The main advantage of wood furnaces is the fan-forced circulation system that allows even heat distribution. All models contain automatic dampers, thermostatic fan control, and safety shutdown dampers. Due to the air-heating plenum design, central air conditioning may be added without difficulty.

Most models of wood-fired hot air furnaces are designed as dual-fuel units. These systems switch from wood to a fossil fuel or electric resistance heat source automatically as the fire dies down.

Before purchasing one of these models ensure that it carries the EPA certification label. Older designs of wood/electric furnaces that were sold during the height of the 1970s oil crisis are woefully inefficient and will eat you out of house and home.

Wood Fuel – The Renewable Choice

Nature provides us with an endless supply of wood. Dead trees rotting on the forest floor produce the same amount of greenhouse gases as if you had burned the wood in your stove. Sustainable woodlot management requires the thinning and cutting of damaged and dying trees. Even in a small acreage, this "waste wood" will provide sufficient fuel for the largest house.

The species and quality of the wood you burn will have a major impact on the ease of use of your heating system. Freshly cut, wet wood can contain up to 50% moisture by weight. Attempting to burn this wet wood will result in difficult ignition, with reduced heat output and greatly increased pollution. On the other hand, properly dried and seasoned wood will ignite rapidly and provide nearly twice the warmth with less work.

Firewood should be cut and split (at least in half) early in the spring and properly stacked. Piles of wood should be neatly arranged in covered rows, allowing space between each successive row for air to blow through. The summer warmth and breezes will quickly reduce the weight of this wood by half, saving your back when you bring it inside next winter and at the same time ensuring that the stored energy is not being used to boil off water and sap in unseasoned wood. During the seasoning process, wood will start to crack and split on the ends and turn a grayish color. Picking up two pieces of well-seasoned wood and banging them together will create a clear ringing tone. Doing this with freshly cut wood will result in a dull "thud."

The type of wood you purchase or cut will also make a difference. Softwoods such as white birch and poplar are fine for fall and spring when you are just a bit chilled. However, winter heating requires serious heating woods such as maple, oak, elm, and ironwood. These woods have a higher density than softwoods, resulting in higher heat output. Pick up an armful of white birch and an armful of maple. The maple feels twice as heavy as the birch, and guess what? It puts out about twice the energy for the same volume. As wood is sold and trucked by volume, purchasing the hardest woods will save you money and reduce the number of trips to the woodshed. Just remember to get at least a little mix of the softer woods for those winter days when the sun is providing some of the energy heating mix.

Purchasing Firewood

Your local dealer will offer you a confusing blend of softwood, mixed wood, mixed hardwood, fireplace cords, face cords and full cords, green or seasoned. It is important to understand what all this means; otherwise your

Figure 3-69. Firewood should be properly stacked and covered and allowed to season for at least six months to drive off moisture. Dry firewood should be a grayish color and have cracks on the ends.

experience with heating in winter may make you think you are back in colonial Pennsylvania.

Firewood is sold in "face cords" or "full cords." The wood is typically delivered pre-cut in 16" (41 cm) logs. When stacked in three rows, a full cord of wood should measure 4' deep x 4' high x 8' long. (1.2m x 1.2m x 2.4 m). A face cord is, as the name implies, one row or "face" of a full cord, which is equal to one-third of a full cord (16" deep x 4' high x 8' long / 0.4m x 1.2m x 2.4 m). But be careful. If the logs of the face cord are cut into 12" (30 cm) lengths, it will yield only a quarter of a full cord.

If firewood is sold "green" it means that the wood is not seasoned. If you can get a discount for buying green firewood and have the time to season it, go right ahead. Wood can also be purchased in 4' or 8' lengths. Provided you have the tools, time, and stamina to cut and split your own wood, the savings may be well worth the effort. To quote Thoreau, "Wood heats you twice, once when you cut it and once again when you burn it."

How Much Wood Do You Need?

The best way to figure this out is to try. A well-insulated house, such as our 3,300 square foot, ultra-insulated home, uses two full cords (4' x 4' x 16'/ 1.2m x 1.2m x 4.8m) of mainly hardwood (and a bit of softwood) per year. A smaller, turn-of-the-century stone home just down the street, which has no insulation to speak of, uses six full cords. It just depends. The best thing to do is to buy more than you think you may require. Covered wood will not rot and should last up to three years, after which it will start to decay. So purchase a bit more and let it sit like money in the bank. You can always use it next year.

3.8
How to Cool a House

Central air conditioning is by far the largest and least efficient load in the home. The best way to keep your home cool is to stop the heat from getting inside in the first place. This might sound simplistic, but a well-insulated home with shading that blocks the summer sun is well protected against overheating. Open windows at night to create a cross-breeze and cool the house. In short, follow all of the guidelines discussed in Chapter 2, "Energy Efficiency", paying particular attention to:

- light-colored roof and wall
- upgraded insulation in walls and roof
- upgraded radiant barrier insulation in attic rafters
- low-emissivity (low-E) windows that are not required to aid in winter heating
- window shading
- nighttime cooling using fans and natural cross-flow ventilation

Evaporative Coolers

For those lucky enough to have relative humidity levels below 30% in the summer, an evaporative cooling unit may be the ticket. These devices are very common in the southwestern region of the U.S. and use very little electrical energy.

Working in the same manner as perspiration cooling our bodies on a hot day, air blown through a wet pad is cooled as the water evaporates. An electric fan draws outside air to this wet pad, humidifying and at the same time cooling it. The conditioned air is blown into the home causing hot, stale air to be expelled outside. Because cooling only works with dry outside air, these systems do not work well where summertime relative humidity exceeds 70%.

Most evaporative coolers are rooftop installations that use a bottom discharge blower to channel conditioned air into the home. Rooftop-mounted units are less expensive than ground-mounted configurations. Evaporative coolers can also be installed as add-ons to conventional refrigeration-type central air conditioning or window-mounted units.

Evaporative coolers require large amounts of moving air to cool a home. For desert climates a unit capable of supplying three to four cubic feet per minute of airflow is required for every square foot of floor area. For most other climates this airflow can be reduced to two to three cubic feet per minute.

Refrigerant-Based Air Conditioning

Most A/C systems are sized far larger than needed. This is done in an effort to make sure that "you're getting what you paid for": a very fast, obvious cooling of the indoor air. It also ensures that you don't call the installer back because the A/C isn't working well. On the other side of the coin, a large unit cools the air but does not have sufficient time to reduce indoor humidity levels. A smaller unit running for a longer period will ensure lower indoor humidity and temperature.

All air conditioners sold in North America must have an EnergyGuide label similar to the one shown in Figure 3-49. Use this label as a guide to compare the energy efficiency of different models. Don't get caught up in a quest for the lowest-cost air conditioner at the expense of energy efficiency. This is false economy, as any first-cost savings will be absorbed by higher operating charges time and time again. These rules apply to window air conditioners as well as central cooling systems.

Room air conditioners are less expensive to operate than central units if you only need to condition one or two rooms. When selecting a room air conditioner, match the area to be cooled to the rated capacity of the unit as shown in the accompanying table.

Area to Be Cooled (Ft²) / (M²)	Capacity (BTU per hour)
100 to 150 / 9.3 to 14	5,000
150 to 250 / 14 to 23	6,000
250 to 300 / 23 to 28	7,000
300 to 350 / 28 to 33	8,000
350 to 400 / 33 to 37	9,000
400 to 450 / 37 to 42	10,000
450 to 550 / 42 to 51	12,000
550 to 700 / 51 to 65	14,000
700 to 1000 / 65 to 93	18,000

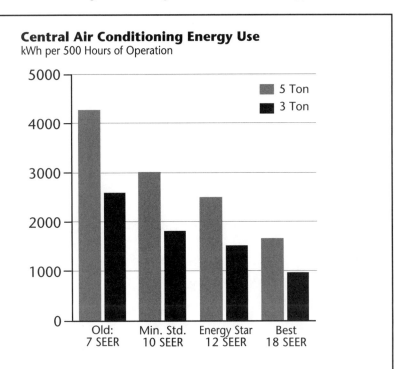

Central Air Conditioning Energy Use
kWh per 500 Hours of Operation

Figure 3-70. Purchasing an energy-efficient air conditioning system will reduce operating costs by more than half. Purchase models carrying the ENERGY STAR label or with a higher efficiency level based on EnergyGuide or Energuide ratings. (Source: Washington State University Extension Energy Program)

4
Community Power

In 1995 the price of natural gas was hovering at around one dollar per million BTUs on the spot market. Anyone heating their home with natural gas during this period will recall their fuel bills being relatively insignificant. In the latter half of 2004, the price of this same "cheap" fuel was around six dollars per million BTUs, inflating nearly 600% in a nine-year period.

Whenever the markets are rattled by price hikes and instability we don't have to look very far to find the reason for the jitters. Sure enough, North American demand for natural gas has skyrocketed in the last decade, and supply has not kept pace. Even Canada, the land of unlimited resources, reached its peak production a few years ago and is now watching supply taper off while demand climbs out of control.

Unfortunately it is not just natural gas supply that is in a state of flux. Oil imports from the Middle East and domestic electrical supplies are all being stretched to the limit. The major grid blackout in the North American Northeast sector that darkened 50 million homes was but one of several major worldwide electrical failures in 2003.

If you're a homeowner or renter the effect is obvious: a bigger chunk of your take-home pay is going to finance the increasingly expensive and difficult exploration for new energy sources, whether they be halfway around the world or right in your own backyard.

There are options, however. As you now know, reducing energy consumption is more economic than increasing supply. Once you have captured all of the energy efficiency options, additional generating sources can then be developed. Why not take part in that investment?

Figure 4-1. The blackout of 2003 that left 50 million North Americans temporarily without power was but one of several major worldwide grid failures that year. Large institutional and industrial consumers are beginning to wonder if they should take matters into their own hands. The advanced technologies of distributed electrical generation have brought this thinking to the community level.

The Case for Community Power

Most people think of energy as coming from centralized infrastructure systems that are extremely capital intensive and best left to governments and major corporations. While that might be true of drilling oil in the North Sea, there is one route the rest of us can take to become involved in society's energy decisions, and that is by building community power.

The traditional electrical supply system has evolved as a natural responsibility of government for reasons of scale and because the stability of such a system requires central control in real time. It developed in its current form, with large generating plants remote from the end user, largely as a result of the distance of energy inputs such as waterfalls, coal seams, and uranium ores from centers of demand. Huge generating plants, with payback periods measured in decades, and the transmission infrastructure necessary to convey that energy across great distances became the norm. The system was logical as long as cost and environmental "externalities" were no object.

Electricity has become an essential public good, in many ways a life-support system provided by government as a cheap and reliable public service. End-users have traditionally been completely passive, able to consume at will without giving a second thought to the act of consumption, let alone to the mammoth tasks of production and delivery.

In addition to cost overruns for generation and transmission, there has been political pressure to keep prices low and uniform regardless of the cost of supply for different regions and without taking into account different classes of consumer. For example, an energy efficient homeowner pays the same rate as their wasteful, power gobbling neighbour. The resulting low tariffs and extensive cross-subsidies have muddied the waters between the provision of electricity and social welfare, fostering considerable inefficiencies in the process. The costs associated with environmental impacts have traditionally been left out of the equation entirely, allowing environmental degradation. However, despite the shortcomings of the traditional model it continues to power society, allowing governments to not question how power systems should be run and therefore becoming deeply entrenched and resistant to change.

The Decentralized Alternative Model

Forms of energy such as landfill gas, wind, biomass, solar thermal, small hydro, and photovoltaics do not have the same problems of scale and location inherent in the centralized model. Newer small-scale or modular technologies developed to harness these inherently clean, natural resources do not have to be remote from centers of demand. Therefore, they do not require extensive infrastructure for power transportation. (Installing a few kilowatts of photovoltaic panels on a residential roof will power that home and possibly the neighbour's, but certainly no more.)

These technologies, along with many innovative methods of improving energy efficiency, have the potential to interest private-

Figure 4-2. One-third of Germany's wind capacity is cooperatively owned by more than 200,000 shareholders holding a market capitalization of over $5 billion. By contrast, North America has only one cooperative wind turbine with 450 investors. (Courtesy Paul Gipe)

sector investors in many different applications, alleviating the scale of the investment otherwise required of the public sector.

Private investment in both efficiency and supply would place much more control in the hands of large consumers or local aggregations of small consumers acting as cooperatives which would no longer need to remain captive or passive. Taken to its logical conclusion, rapid technological innovation combined with policy incentives to implement alternative energy generation choices is providing the means for society to implement decentralized generating capacity and ownership programs.

Approximately one-third of Germany's 15,000 MW of installed wind capacity is cooperatively owned by more than 200,000 private shareholders representing a market capitalization of over $5 billion. By contrast, North America has only one cooperative wind turbine venture with 450 local investors.

Figure 4-3. Cooperative wind turbines such as these units located in Noordoostpolder, The Netherlands, decentralize ownership and help distribute wealth as opposed to concentrating it in the hands of a few industrial conglomerates. (Courtesy Paul Gipe)

Cooperative ownership of energy generation infrastructure has multiple benefits such as providing a return on investment which is directly related to the energy efficiency of the cooperative. Members who reduce their own energy consumption not only pay less for their power but also contribute to generating capacity that may be sold profitably to a third party. In essence the owner of shares in the cooperative has a direct incentive to become energy efficient.

Europe has higher population density and larger resource constraints than North America, which has given it a head start in the sustainable energy arena. However, rising energy costs, environmental concerns, and advances in distributed generation technologies are all working in harmony to help North Americans play catch up.

Figure 4-4. These shareholders at a wind cooperative meeting in Copenhagen, Denmark, receive financial dividends from the venture as well as education relating to energy generation. Having a direct incentive to reduce consumption to increase their dividend payment makes these Danes some of the most energy-efficient people on the planet. (Courtesy Paul Gipe)

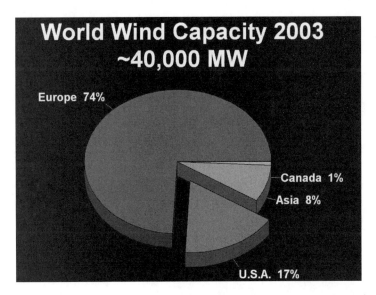

Figure 4-5. Installed world wind capacity was approximately 40,000 MW as of 2003. With 74% of this capacity, Europe leads the way in distributed and cooperative generating systems. By contrast, Canada has less than 1% of this total or approximately 400 MW capacity, even though Canadians are the largest per capita energy consumers. (Courtesy Paul Gipe).

The distributed generation and community power model does not require people to invest their life savings in wind turbines or cooperative ventures. Even the seemingly simple things such as public outreach, shopping locally, and using mass transit help to make society and the local community more sustainable.

Community power involves more than merely installing wind turbines and issuing shares to homeowners. Tangible benefits to the local community and society at large result from these programs. Community power is locally owned and situated. It is clean, sustainable energy that economically benefits the surrounding region. Community power entails medium- to large-scale generating equipment in small, economically feasible instalments.

Public Outreach

Possibly the most important influence on society to adopt energy efficiency and sustainable energy programs is education. Schoolchildren know more about calculus and Chaucer than they do about kilowatts and conservation, which is odd given that a large portion of people's income is siphoned away paying for energy on an ongoing basis. Even at today's relatively low energy rates, a young homeowner can expect to pay several thousand dollars per year fuelling cars, heating homes, and keeping the lights on. Given that this expense continues for decades, and without factoring rising costs, the average family can easily expect to pay in excess of a quarter of a million dollars

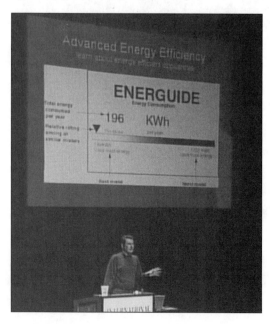

Figure 4-6. The author lecturing a group of concerned citizens on the need for energy efficiency in the home. Many people don't realize that the money they're spending on higher-than-necessary utility bills could be used to purchase energy-efficient appliances. The money spent on heating water with electricity for six years would easily cover the cost of a solar water heater. (Courtesy Private Power Magazine)

on energy in a lifetime.

It doesn't have to be this way. Almost every city, town, and village in North America has either an energy efficiency program or "eco-coalition" working to make the local environment more sustainable. Many of these groups sponsor lectures (Figure 4-6) and renewable energy fairs (Figure 4-7) or offer "Green Living" workshops and training programs similar to those developed by the Real Goods Solar Living Institute (Figure 4-8).

If education related to saving energy sounds about as much fun as having a root canal, it needn't be. Strolling through a fair or tradeshow while learning about environmental and energy technologies that save money and the earth, is a pleasant way to spend an afternoon. After all, you could be studying calculus with the kids.

Figure 4-7. Many North American universities develop solar-powered race cars which showcase their engineering finesse. Throughout the summer, teams of students race these vehicles across continents vying for prestige and raising public awareness. Looking like Darth Vader's helmet on wheels, these vehicles are always crowd pleasers.

Local Dealers

Education is the engine that gets the renewable energy wagon rolling, but you still need a dealership to sell the wagon in the first place. There are thousands of technologies designed to save or generate energy that "guarantee" an excellent return on investment. Sadly, the guarantee isn't always valid. Although most dealers are honest, reputable people who are an integral part of the community and recognize that fleecing their clients is not in their long-term best interests, others aren't quite as principled. And even if the

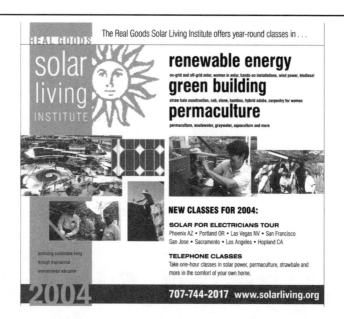

Figure 4-8. The Real Goods Solar Living Institute located in Hopland, California, was one of the first organizations to offer classes in sustainable living and renewable energy. Now that these technologies are mainstream, numerous programs are springing up offering education from beginner level to expert. (Courtesy Solar Living Institute)

dealer is doing a respectable job, competing technologies can make the purchasing process more complex, requiring expert and unbiased advice.

When faced with the prospect of spending money, whether it is a few dollars or a few thousand dollars, seek the advice of a reputable dealer. Local retailers that have been in business for a number of years and provide a clean, neat storefront are likely to provide honest and unbiased service. As you would with any sizeable purchase, consider obtaining references from past clients as well as visiting installations to see exactly how well the product operates in the real world.

Energy efficiency products (such as compact fluorescent lamps and low-flow shower heads) tend to be relatively inexpensive and may be purchased through local environmental or hardware stores. Renewable energy generation equipment is often "co-marketed" with complementary products such as wood stoves, swimming pool and spa products, or heating and air-conditioning equipment.

Just like every other retail product in North America, renewable energy systems may be purchased via the Internet from reputable dealers and used

Figures 4-9 and 4-10. Environmental stores offer a wide array of ecologically responsible products including energy efficiency devices. Arbour Environmental Shoppe located in Ottawa, Canada, has a perfect mix of useful products to help lessen your environmental impact.

Figure 4-11. Solar-powered radios, energy-auditing software, and photovoltaic-powered toys are but a few of the resources and educational products carried by many environmental outlets. Some of these products are kits which make great gifts and can even distract kids from the television for a few hours.

Figure 4-12. Renewable energy generation equipment is often "co-marketed" with complementary products such as wood stoves, swimming pools, or heating and air-conditioning equipment.

equipment traders or through auctions on web sites such as eBay. Unless you are an experienced home handyman (or woman) it is wise to stick with local dealers you can trust. While it may be possible to have an inverter serviced over the phone by someone at a call center a thousand miles away, having a friendly supplier just down the street is very comforting.

Many dealers not only sell the equipment but offer *qualified* installation and repair service. Warming hot water with the sun sounds like child's play until you examine the technology required to harvest the sun's energy and transfer it safely and reliably into your water tank. Differential temperature controls, expansion tanks, vacuum solar collectors, and a myriad of other bits and pieces must be carefully orchestrated and installed, requiring quality workmanship to ensure a long and trouble-free operating life. Many jurisdictions require licensed contractors to perform the installation work, demanding building permits and insurance certificates prior to completion.

Dealers are more than warehouses for the products they sell. They are experienced, professional, and educated people who want to be sure that you are satisfied with their products. While you may be able to purchase goods at a lower price, quality dealers are worth their weight in gold through the installation, operation, and lifelong servicing of their products.

Figure 4-13. When shopping for a renewable energy products dealer, look for a store that is well maintained and well stocked. Get references from current and past customers and try to arrange a site visit with a previous client.

Living in the Community

Cutting the grass and taking out the garbage isn't everyone's idea of Utopia. In larger and more expensive urban centers apartments and condominiums may better suit your needs. From an energy efficiency, land use, and transportation standpoint, high-density housing beats urban sprawl hands down.

Located in Santa Monica, California, Colorado Court was designed to exploit passive environmental control strategies such as natural ventilation, maximized daylight, and shaded south-facing windows. It also incorporates a number of innovative energy-generation measures, notably a natural gas powered turbine and heat recovery system that generates electricity and uses waste heat to service the building's hot water requirements. The photovoltaic panels seen in Figures 4-14 and 4-15 supply most of the peak load energy demand and feed unused energy into the local electrical grid at a profit. The designers, Pugh + Scarpa Architecture, estimate that these systems will pay for themselves in less than ten years. Annual savings in electricity and natural gas of approximately $6,000 help to keep rents between $316 and $365 a month.

At the other end of the monthly rent scale is the ultra-efficient sky scaper known as the Conde Nast building in New York City. This commercial building is the heart and sole of our economic prosperity and a place

Figures 4-14 and 4-15. Colorado Court, located in Santa Monica, California, is a five-story, forty-four unit, single room occupancy apartment complex for low-income tenants. It is one of the first buildings of its kind in the United States that is 100% energy independent, generating nearly all its own energy for electricity, heat, and light. (Courtesy Pugh + Scarpa Architecture)

where millions of North Americans spend the majority of their working day. Sustainable buildings should work with nature, allowing artificial light to harmonize with natural light sources. Integrating efficient and automatic lighting reduces energy losses, waste heat and increased demand on air-conditioning systems. These and other state-of-the-art improvements in design and efficiency result in astounding utility cost savings.

High-performance buildings such as the Conde Naste facility should be energy efficient and have a low impact on the environment as well as being healthy and enjoyable workplaces. Using building certification programs such as the Leadership in Energy and Environmental Design (LEED) program can easily reduce energy consumption and associated costs by 50% without

Figure 4-16. The Conde Nast building located at 4 Times Square in New York City is an indication of where urban energy-efficient buildings with integrated/renewable energy generation technologies are heading. This is one of the few buildings that remained lit during the 2003 northeast power blackout. (Courtesy Fox & Fowle Architects / Andrew Gordon Photography)

increasing capital costs. It sounds almost too good to be true, so why don't more buildings employ these technologies?

The problem with many commercial projects is a near-sighted focus on first-time costs while leaving operating costs for the tenant or building owner to worry about. There is a disconnection between the builder's concern for first cost and the tenant's worry about operating cost that has not been successfully bridged. Energy efficiency standards including upgraded building codes are the keys to enabling widespread application of these technologies.

One can get a sense of what's coming by looking at the Conde Nast building during the August 2003 blackout. You may recall that this building was one of the few that remained lit. This was possible because of two fuel cells installed as part of a distributed energy system designed to meet 100% of night time electrical demand.

Figure 4-17. High-performance building design is promoted by the Leadership in Energy and Environmental Design (LEED) certification program, which ensures that buildings are energy efficient as well as healthy and enjoyable places to work and live.

Figure 4-18. Mass transit using hybrid vehicles, electric trains, and biodiesel fuel reduces smog and greenhouse gas emissions. Taking a train or bus decreases the amount of fuel required per passenger mile dramatically (and saves you exorbitant parking fees).
(Courtesy Saskatchewan Canola Development Commission)

During normal operation, photovoltaic panels installed on the south and west sides of the building feed inverter driven AC power into the local utility grid for sale. Coupled with advanced energy-management controls, enhanced insulation and windows, energy-efficient lighting, and occupancy sensors, the Conde Naste building exceeds New York Building Code requirements by 35%.

With mass transit powered by hybrid vehicles, electric trains, and biodiesel buses, urban smog may become a thing of the past. Perhaps the future of urban living is looking brighter after all.

Generating Power in the Community

Renewable energy is inherently decentralized. Farmers, cooperatives, rural communities, and First Nations properties are ideally located to install these systems, due to their proximity to natural energy sources such as rivers and windy tracts of land. In Germany, renewable energy tariffs have led to individuals investing $1.7 billion and installing 20,000 solar-electric systems in one year. Since 1991, the country has also installed 14,000 MW of wind generation, a renewable energy boom by any standards. Photovoltaics in particular make their maximum contribution during the summer consumption peak which defines the required capacity of the electrical generating system. Installed in sufficient quantities they can reduce the need for expensive peak generation plants or power imports.

Community-based power systems have significantly aided the growth of the renewable energy industry in Europe, with wind power providing approximately 20% of the electrical needs of both Denmark and Germany. The community and cooperative ownership model has been key to this successful implementation. The countries of Denmark and Germany currently

Figure 4-19. The Toronto Renewable Energy Cooperative (TREC) wind turbine located on the edge of Toronto, Canada, is North America's first urban, cooperative wind-based project. A 50% share of the turbine is held by 450 private shareholders. (Courtesy Toronto Renewable Energy Cooperative)

have over 300,000 households which own shares in a local wind power cooperative. In stark contrast stands the United Kingdom, where flawed energy policies have favored the centralized model with distributed generation floundering to insignificant levels.

Community power is more than just installing wind turbines and issuing shares to homeowners. Tangible benefits to the local community and society at large result from these programs. Community power is locally owned and situated. It is clean, sustainable energy that economically benefits the surrounding region. Community power revolves around medium- to large-scale generating equipment installed in small, economically feasible sizes.

The benefits of community power include:

- improved economic development featuring long-term, skilled employment;
- new income strategies for farmers and rural landowners;
- positive support for proposed projects helping to eliminate the "not in my backyard" phenomenon;
- reduced electrical losses and decreased need for transmission system upgrades resulting from generation of electricity close to the natural energy source and demand centers;
- increased energy efficiency as a result of education and outreach related to the project, with the community understanding where energy comes from.

Most community power projects involve a partnership between a community-based cooperative or corporation and a local developer, municipality, or utility. Bringing community-based projects to life is by no means simple, requiring the organizing body to clear regulatory and financial hurdles. Communities and municipalities looking to enter the power generation business are advised to discuss project details with other groups who have completed projects of a similar scale.

Figure 4-20. This miniature jet engine or microturbine is part of a combined heat and power (CHP) system which uses natural gas to produce electricity for sale to the electrical utility while capturing waste heat to provide heat and hot water. These units are very efficient and will become common in cooperative and condominium housing complexes. (Courtesy Mariah Energy Inc.)

Figure 4-21. Small hydro projects such as this 500-kilowatt facility are ideal for small cooperatives and communities. Properly designed and installed systems cause negligible environmental interference for marine life or the surrounding ecosystem.

Figure 4-22. The early 1900s ushered in the era of small hydro power plants in most communities, but they faded away in favour of large-scale centralized power generation. With advances in turbine and electronic control technologies, small-scale hydro is beginning a renaissance.

5
Biofuels

Humans have very short memories. If you were of driving age in 1973 you will no doubt recall the long lineups and short supply of gasoline at your local filling station. Even if gasoline was available, OPEC raised the price of oil from $4.90 a barrel to $8.25 a barrel in that year. None of this had anything to do with diminishing world oil supplies, but came about as a result of the American support of Israel and the misguided energy policies of the Nixon administration.

Figure 5-1. Misguided energy policies of the Nixon administration coupled with the Arab oil embargo over American support of Israel led to the severe oil supply shortages in the early 1970s. (Courtesy U.S. National Archives and Record Administration)

Energy consumption in the United States and Canada is insatiable, and it appears no one has learned from the mistakes of the 70s. Look around: Sport Utility Vehicles (SUVs) and minivans abound, with the result that fuel economy is nearing an all-time low. At the same time, the square footage of the average North American home has more than doubled, requiring additional energy for heating and air conditioning.

Over the same period domestic oil supplies in the United States have dropped to an insignificant amount of total demand, requiring vast quantities of imported oil to make up the deficit. Current world consumption is around eighty million barrels of oil per day and rising quickly because developing countries are demanding their share of the energy pie. Even though the majority of the world's oil is supplied by regions that are politically unstable and often unfriendly to the West, energy dollars are being exported as fast as the U.S. Treasury Department can print them. According to an October 25, 2003 report in *The Economist*, OPEC has drained the staggering sum of $7 trillion from American consumers over the past three decades. This massive sum does not even include industry subsidies, cheap access to government land for oil extraction, or military security required to get the sticky stuff safely into North America. If this money had been pumped into solar energy and fuel cell and hydrogen technologies, the internal combustion engine would be as common as a woolly mammoth.

Many North Americans are astounded at the cost of gasoline in Europe. It is interesting to note that the price of oil (and hence gasoline) is a world price. European countries decided long ago that taxing the daylights out of transportation fuel would improve energy efficiency and reduce both the size of motor vehicles and the number of needless miles driven. As prices

Figure 5-2. The United States has put itself in the very difficult position of having to rely almost exclusively on imported oil to keep its economy running. Unlike European governments, decades of North American administrations have not had the backbone to improve the Corporate Average Fuel Economy (CAFE) laws or to tax gasoline in order to reduce consumption.

nudge $50 per barrel, up from $11 in 1998, many people will begin to question the wisdom of $80 fill-ups every couple of days and the need for cars as large as my parents' first house.

On the other hand, gasoline prices in the United States are virtually untaxed and don't include the direct costs of military muscle in the Middle East or damage to the environment. The after-inflation price of gasoline has actually dropped since the oil crisis of a couple of generations ago, to the point where a quart of bottled water costs more than a quart of oil.

Figure 5-3. Oil prices in the United States are virtually untaxed and don't include the costs of military muscle in the Middle East or damage to the environment. The after-inflation price of gasoline has actually dropped since the oil crisis of thirty years ago, to the point where a quart of bottled water costs more than a quart of oil.

Reducing consumption by increasing efficiency is a fairly easy fix (provided you're prepared to do it) and is covered in Chapter 6, "Energy-Efficient Transportation". If you must keep your beloved Hummer or, for that matter, your fuel-efficient Toyota running, gasoline and diesel fuel will continue to power The American Dream.

An Alternative to Fossil Fuel

While the world waits breathlessly for hydrogen-powered fuel cells and advanced electric vehicles, developments in biofuels are continuing at a surprising rate. Almost everyone has heard of ethanol and some enlightened souls may be familiar with biodiesel. While these technologies are not nearly as sexy as hydrogen and fuel cells, they are available now and offer numerous advantages over imported oil.

Modern internal combustion engines including those used in transportation vehicles do not have to run on gasoline. In fact, early automotive pioneers did not have access to refined gasoline and used peanut oil and alcohol for fuel. Coincident with the development of the internal combustion engine was the discovery of large amounts of crude oil in the United States. With ready access to this low-cost energy source, the days of the gasoline-powered buggy were upon us.

Fossil fuels in the form of coal, oil, and natural gas originated eons ago as plant matter and marine plankton growing in the ancient world. All living plants use the sun's energy in a process known as photosynthesis to convert atmospheric carbon dioxide into carbon which is stored in the plant structure. The modern fireplace or woodstove can burn firewood and convert the stored carbon back to heat and atmospheric carbon dioxide.

When a plant dies and becomes trapped in mud or bogs, decomposition is stalled and under conditions of high heat and pressure fossilization takes place, forming coal. Marine plankton follows a similar fate, converting into oil and natural gas.

5.1 Ethanol-Blended Fuels

Fossilization is not the only means of extracting energy from plant life. Fermentation of grapes and apples has been fuelling alcohol-induced binges for as long as man can remember. The naturally occurring sugars in the fruit produce wine and cider with a maximum alcohol content of approximately 12%. Applying heat to these beverages, in a process known as distillation, allows extraction of the alcohol at up to 100% concentration.

The primary source of plant sugars can vary, as automotive-grade ethanol can be produced from grains such as corn, wheat, and barley. Recent advances in enzymatic processes even allow the conversion of plant waste in the form of straw and agricultural residue into sugars which can in turn be fermented into ethanol.

Conventional grain-derived and cellulose-based ethanols are the same product and can be easily integrated into the existing gasoline supply chain. Ethanol may be used as a blending agent or as the main fuel source. Gasoline blends with up to 10% ethanol can be used in any vehicle manufactured after 1977. High-level ethanol concentrations of between 60% and 85% can be used in special "flex-fuel vehicles." At the time of writing Ford, General Motors, and DaimlerChrysler warranties allow up to 10% ethanol blends in their North American vehicles.

Adding ethanol to gasoline increases octane, reducing engine knock and providing cleaner and more complete combustion, which is good for the environment. Compared with gasoline, ethanol reduces greenhouse gas emissions: a 10% ethanol blend with gasoline (known as E10) will reduce GHG emissions by 4% for grain-produced ethanol and 8% for cellulose-based feed stocks. At concentrations of E85, GHG emissions are reduced by up to 80%.

Greenhouse gas emission reduction was abundantly demonstrated during a recent 6,000-mile (10,000 km) driving tour conducted b y Iogen Corporation. An SUV fuelled with 85% cellulose ethanol produced the same GHG emissions as a super-efficient hybrid car (see Chapter 6).

In addition to making environmental improvements, ethanol is produced from domestic renewable agricultural resources, thereby reducing our dependence on imported oil. Even at low ethanol/gasoline concentration levels, ethanol production is a major source of economic diversity for rural farming economies.

In the United States, 77 ethanol plants produce over 3.3 billion gallons (12.5 billion liters) of ethanol per year. Canadian production currently stands at 62.6 million U.S. gallons (237 million liters) per year. Of the 77 plants in

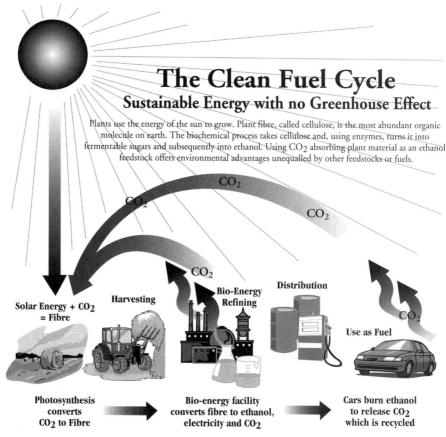

The Clean Fuel Cycle
Sustainable Energy with no Greenhouse Effect

Plants use the energy of the sun to grow. Plant fibre, called cellulose, is the most abundant organic molecule on earth. The biochemical process takes cellulose and, using enzymes, turns it into fermentable sugars and subsequently into ethanol. Using CO_2 absorbing plant material as an ethanol feedstock offers environmental advantages unequalled by other feedstocks or fuels.

CO_2

CO_2

CO_2

CO_2

CO_2

Distribution

Bio-Energy Refining

Harvesting

Solar Energy + CO_2 = Fibre

Use as Fuel

| Photosynthesis converts CO_2 to Fibre | Bio-energy facility converts fibre to ethanol, electricity and CO_2 | Cars burn ethanol to release CO_2 which is recycled |

Figure 5-4. Plant grains and fibre can be converted to sugar, fermented into ethyl alcohol (ethanol), and used as a blending ingredient with gasoline or as the main fuel. High concentrations of ethanol can reduce greenhouse gas emissions by up to 80% relative to gasoline. (Courtesy Iogen Corporation)

Figure 5-5. The majority of ethanol is derived from corn. The production of ethanol fuels strengthens agricultural regions and creates new and more stable markets for the farming community.

the United States, 62 use corn as the feedstock. The remainder use a variety of seed corn, corn and barley, corn and beverage waste, cheese whey, brewery waste, corn and wheat starch, sugars, corn and milo, and potato waste. In the United States there are 55 proposed new plants and 11 currently under construction. Canada currently has 6 plants with 1 under construction and 8 new proposals on the drawing board.

There is no question that using domestically grown feed stocks from renewable and clean-burning grains is better than importing fossil fuel from the Middle East. However there is some question about diverting food stocks into fuel feed stocks to continue our love affair with the automobile, thereby contributing to urban sprawl and other societal problems. A reduction in automotive usage is a key, long-term goal but one fraught with difficulty in the short term. Until urban planners can wrench the car keys from suburbanites, society might as well use all the clean "transition" fuels at its disposal.

Cellulose ethanol eliminates the diversion of food crop to fuel feed stocks and is an advanced new transportation fuel which has some advantages over grain-based ethanol:

Figure 5-6. North American ethanol production currently stands at 3.36 billion gallons (12.72 billion liters) and is expected to increase by 28% before the end of the decade. This is just a drop in the bucket compared with fossil-fuel gasoline usage, but with good demand management (through CAFE and properly managed taxation) it could go a long way toward reducing reliance on imported oil. (Courtesy Iogen Corporation)

- Unlike grain-based ethanol production, the manufacturing process does not consume fossil fuels for distillation, further reducing green house gas emissions.
- Cellulose ethanol is derived from non-food renewable sources such as straw and corn stover.
- There is a potential for large-scale production since it is made from agricultural residues which are produced in large quantities and would otherwise be destroyed by burning.

Figure 5-7. Adding ethanol to gasoline increases octane, reducing engine knock and providing cleaner and more complete combustion. In addition to performance and environmental improvements, ethanol is produced from domestic renewable agricultural resources, reducing dependence on imported oil and providing a major source of economic diversity for rural farming economies.

Cellulose-based ethanol has, in the past, been very expensive as a result of the inefficient processes required to produce it. The Iogen Corporation, working in conjunction with the Government of Canada and Shell Corporation, has recently launched a cost-effective method for producing what the company refers to as EcoEthanol™. Once the first phase of production, using straw and corn stover, is under way, it won't be long before other feed stocks can be used, including, hay, fast-growing switch grass, and wood-processing byproducts.

Ethanol Considerations

There are very few downsides to ethanol fuel. Because ethanol contains oxygen it permits cleaner and more complete combustion which helps keep fuel injection systems deposit free. Be aware, though, that ethanol acts as a solvent and can loosen residues in your car's fuel system, necessitating more frequent fuel filter changes.

Gasoline containing 10% ethanol has approximately 3% less energy than regular gasoline. However, this loss of energy is partially offset by the increased combustion efficiency resulting from the higher octane and oxygen concentration in the ethanol blend. Actual road tests show that fuel economy may be reduced by approximately 2% when using E10 concentrations. To put this in perspective, driving 12 mph (20 kph) over the 60 mph (100 kph) speed limit increases fuel consumption by an average of 20%.

Lastly, there may be a slight cost increase for ethanol-blended gasoline in your area. However, compared to the long-term alternatives of the fossil-fuel market, injecting a few pennies per gallon back into the domestic rural economy and helping the environment at the same time is surely the right thing to do.

Figure 5-8. Iogen Corporation President Brian Foody stands in a field of straw grass, which may become the twenty-first century equivalent of a Saudi Arabian oil field. (Courtesy Iogen Corporation)

5.2
Biodiesel as a Source of Green Fuel and Heat

Under conditions of extreme heat and pressure, marine plankton are transformed over millennia into crude oil and natural gas. Refining crude oil through a cracking and distillation process produces gasoline, kerosene, diesel fuel, and other hydrocarbon compounds.

Internal combustion engines ignite fuel using one of two methods. A spark ignition engine produces power through the combustion of the gasoline and air mixture contained within the cylinders. An electric spark which jumps across the gap of an electrode ignites this volatile mixture.

Figure 5-9. Loading straw into an ethanol plant is a lot cleaner and much more sustainable than extracting oil from the North Sea. Waste grasses, straw, and wood products are "carbon-neutral" fuels that are available domestically and in endless supply. Think of these sources as "stored sunshine." (Courtesy Iogen Corporation)

The compression ignition or diesel engine, as it is more commonly known, uses heat developed during the compression cycle. (Have you ever noticed how hot a bicycle air pump becomes after a few strokes?) With a compression ratio of 18:1 or higher, sufficient heat is developed to cause diesel fuel sprayed into the cylinder to self-ignite. The higher energy content of diesel fuel as compared with gasoline contributes to improved fuel economy and power, desirable features for fleet and other high-mileage vehicles.

Biodiesel, like its cousin ethanol, is a domestic, clean-burning renewable fuel source for diesel engines and oil-based heating appliances. It is derived from virgin or recycled vegetable oils and animal fat residues from rendering and fish-processing facilities. Biodiesel is produced by chemically reacting these vegetable oils or animal fats with alcohol and a catalyst to produce compounds known as methyl esters (biodiesel) and the by-product glycerine.

Biodiesel that is destined for transportation fuel must meet the requirements of the American Society of Testing and Materials (ASTM) Standard D6751 for pure or "neat" fuel graded B100. Fossil fuel diesel (or petrodiesel) must meet its own similar requirements within the ASTM standard. Biodiesel and petrodiesel can be blended at any desired rate, with a blend of 5% biodiesel and 95% petrodiesel denoted as B5, for example. The testing and certification of any transportation fuel is a requirement of automotive and engine manufacturers implemented to minimize the risk of damage and related warranty costs.

Figure 5-10. Biodiesel fuel is a product of vegetable oils or waste animal fats which produce a renewable, clean-burning fuel source for diesel engines and oil-based heating appliances.

From an agricultural viewpoint, biodiesel offers many of the same advantages to the farming community as ethanol feed stocks. In addition, the process of making biodiesel is *relatively* simple and low cost, which may lead to cooperative and rural ownership of processing and production facilities, further increasing farming income and risk diversity. Even the by-product of the biodiesel manufacturing process, glycerine, has a ready market in the food, cosmetic, and other industries.

By way of example, Milligan Bio-Tech Inc. located in rural Saskatchewan, Canada is a community-owned agri-business venture. The canola seed growers in the region produce a food-grade, crystal clear canola oil for the food industry. Vagaries in the weather may cause the oil seed to produce off-color oil that is rejected by the market. Crop failures are all too common in farming communities, resulting in financial hardship or worse. Progressive thinking in the community led to the creation of Milligan as a way of diversifying crop development and reducing the economic risk inherent in the canola business.

Milligan realised that simply selling the basic seed and food oil would not provide sufficient diversification for the business. Assessing its options, Milligan determined that a mobile seed crushing/oil extraction plant would allow the seed husk to be left at the farming site, providing oil seed meal that could be used as a feed supplement for livestock.

Virgin oils could then be graded and sold as food oil or placed into the biofuel and lubricant processing stream. As discussed earlier, vegetable oil produces biodiesel and the by-product glycerine. Biodiesel may be sold directly as a transportation fuel or refined into penetrating or lubrication oil. It can also be used as a carrier in diesel fuel additives which reduce emissions and increase lubricity, prolonging engine life. Milligan has even developed a novel use for the glycerine by-product, creating an ecologically friendly, biodegradable, and long-lasting dust suppressant for gravel roads which can replace salt-based calcium chloride products currently in use.

According to Milligan principal Zenneth Faye, rural communities cannot afford to be dependent on one income stream without taking unnecessary financial risks. Diversification is the key to community growth and sustainability.

Biodiesel Performance

The modern diesel engine is a far cry from the smoky, anaemic model of the 1970s. As a direct result of Mr. Nixon's oil crisis, consumers lined up to purchase Volkswagen Rabbit and Mercedes diesel cars, enticed by fuel economy claims. In addition to being disillusioned by the lack of power and acceleration, when the mercury dipped below 32°F (0°C) and a diesel engine wasn't plugged in overnight a bus ride was a sure bet the next morning. Is it any wonder that people lift their noses at diesel-powered cars?

Figure 5-11. As the price of petrodiesel continues to rise and biodiesel production costs fall, rural communities will produce their own democratic energy to fuel their part of the economy.(Courtesy Lyle Estill/Piedmont Biofuels)

Fast forward to today. Gone are the smelly, smoky, lumbering diesels of old. Witness the new Mercedes E320 family of "common rail, turbo-diesel" engines that offer no "dieseling" noise, smoke, or vibration, achieve superb mileage, and have better acceleration than the same model car equipped with a gasoline engine.

Biodiesel offers some distinct advantages as an automotive fuel:

• It can be substituted (according to vehicle manufacturer blending limits) for diesel fuel in all modern automobiles. B100 may cause failure of fuel system components such as hoses, o-rings, and gaskets that are made with natural rubber. Most manufacturers stopped using natural rubber in favour of synthetic materials in the early 1990s. According to the U.S. Department of Energy, B20 blends minimize all of these problems. If in doubt, check with your vehicle manufacturer to ensure compliance with warranty and reliability issues.

• Performance is not compromised using biodiesel. According to a 3 1/2 year test conducted by the U.S. Department of Energy in 1998, using

Figure 5-12. The simplicity of a biodiesel production facility is shown in this aerial view of a continuous production plant. Soybeans are delivered to the processing plant for conversion to food-grade oil. Alcohol and a catalyst are added to the oil to produce biodiesel and glycerine. The biodiesel is stored in the tank farm. The glycerine byproduct is sold while the water and alcohol are recycled, creating an environmentally friendly processing cycle. (Courtesy West Central Soy)

low blends of canola-based biodiesel provides a small increase in fuel economy. Numerous lab and field trials have shown that biodiesel offers the same horsepower, torque, and haulage rates as petrodiesel.

- Lubricity (the capacity to reduce engine wear from friction) is considerably higher with biodiesel. Even at very low concentration levels, lubricity is markedly improved. Reductions in the sulphur levels of petrodiesel to meet new, stringent emissions regulations have, at the same time, reduced lubricity levels in petrodiesel dramatically. Biodiesel blending is currently being considered by the petrodiesel industry as a means of circumventing this problem.
- Because of biodiesel's higher cetane rating, engine noise and ignition knocking (the broken motor sound when older diesels are idling) is reduced.

Figure 5-13. Over 200 public and private fleets in the United States and Canada currently use biodiesel, and the number is increasing rapidly. Environmental stewardship regarding climate change, urban smog, and air quality is creating mass-market acceptance of biodiesel fuel. This bus equipped with a bike rack, allows commuters to use a combination of public transit and peddle power for their commute. (Courtesy Kingston Transit)

Cold Weather Properties

Automotive diesel is supplied as low-sulphur No. 1 or No. 2 grade. Home heating fuel is similarly graded, although it contains higher levels of sulphur. (It may come as a surprise to many, but home heating oil is essentially the same stuff that is burned in road vehicles). No. 2 grade is the preferred fuel as it has a higher energy content per unit volume, giving better economy whether powering a truck or heating a house. Unfortunately, No. 2 is subject to an increase in viscosity known as "gelling" when temperatures drop.

The cold weather performance of biodiesel is similar to that of petrodiesel No. 2, although its pour point temperature (the temperature at which it gels) is slightly better than No. 2 at -11° F (-24° C). Since No.1 has a pour point of approximately -40° F (-40° C), fuel suppliers blend No.1 and No.2. diesel in ratios appropriate for the geographical location and time of year. To prevent gelling of No. 2 petrodiesel or biodiesel blends, it is recommended to store the fuel in an underground or indoor tank should the fuel be subjected to extremely cold weather conditions. Alternatively, your fuel supplier may be able to blend the appropriate mix of No.1 and No. 2 and biodiesel fuels for outdoor storage.

World Energy Alternatives, LLC conducted a number of tests on cold weather performance in the winter of 2002-2003. Working with distributors throughout the United States, the company delivered B20 blended fuel to Malmstrom Air Force Base in Montana, Warren AFB in Wyoming, Peterson AFB in Colorado Springs, and the Winter X Games in Aspen. World Energy biodiesel was introduced to Toronto Hydro in 2001, the first Canadian com-

Figure 5-14. World Energy Biodiesel is delivered to Aspen ski area to fuel the generators and snowcats at Buttermilk Mountain. Numerous ski and eco-sensitive areas are economically and responsibly using biodiesel to meet greenhouse gas emission targets and promote sustainable fuel options. (Courtesy World Energy Alternatives LLC.)

mercial fleet to use the fuel to reduce urban smog. Biodiesel is also used to fuel the generators and snowcats at Buttermilk Mountain ski resort in Aspen, Colorado. World Energy indicates that there hasn't been a single complaint or operational problem with their biodiesel fuel.

Rev Up Your Furnace

Heating fuel oil consumption in the United States is currently hovering around 7 billion gallons (26.5 billion liters) per annum according to the U.S. Department of Energy, with the vast majority supplied by imported oil. Blending a mixture of 20% biodiesel with No.2 would reduce fossil fuel consumption and replace that amount of petrodiesel with domestic, renewable, clean-burning biodiesel. If this strategy were adopted across the United States, biodiesel consumption would increase by 1.4 billion gallons per year (5.3 billion liters per year), reducing carbon monoxide, hydrocarbon, and particulate emissions by approximately 20% while supporting domestic, agricultural economies.

Are you ready to start using biodiesel? The downside is that finding biodiesel can be challenging. In the Northeastern U.S., suppliers are starting to fill this niche market, albeit slowly. Many distributors purchase biodiesel from large producers such as World Energy Alternatives LLC (www.worldenergy.com) or enter a purchasing (and often producing) cooperative such as Piedmont Biofuels of Chatham County, North Carolina

Figure 5-15. Biodiesel comes to the masses. The author (seen grinning enthusiastically, rear) was the first Canadian to fill up with commercially produced biodiesel at this pumping station in Toronto, Canada, in March 2003. Although North Americans are just beginning to use biodiesel, Europeans have been filling up on it for some time now.

Figure 5-16. Blending a mixture of B20 (20% biodiesel to 80% petrodiesel) would reduce carbon monoxide, hydrocarbon, and particulate emissions by approximately 20%. In addition, biodiesel would assist rural, domestic suppliers and help keep North American money at home. (Courtesy West Central Soy)

(www.biofuels.coop). According to Piedmont partner Lyle Estill, production limitations and price can become a deterrent to many people. Piedmont cannot keep up with demand even though it charges $3.50 per gallon ($0.92 per liter) for B100. Of course, if you purchase a B20 blend the price difference compared to petrodiesel is quite small while the environmental improvements are considerable.

For information regarding supply of biodiesel consult the National Biodiesel Board in the United States at www.biodiesel.org (1-800-841-5849) and in Canada the Biodiesel Association of Canada at www.biodiesel-canada.org.

Biodiesel acts as a weak solvent which will soften natural rubber hoses, seals, and gaskets over time. If your furnace was manufactured after 1990 it is likely that these items are made with artificial rubbers such as neoprene which are not damaged by the solvent action. If you are uncertain, contact a furnace service company for advice.

The solvent action will also loosen deposits of "crud and sludge" that are in the tank and fuel lines. If there is a large amount of this build-up present, it will dislodge and eventually plug the fuel filter. For this reason, studies recommend that you switch to the desired biodiesel blend level (i.e. B20) immediately, rather than building up the concentration over time. This allows the deposits to be released quickly, requiring only one or two fuel filter changes rather than having to repeat the process over and over.

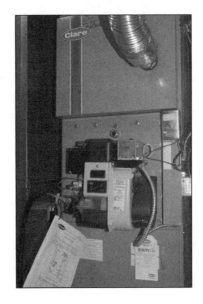

Figure 5-17. Biodiesel can be used in virtually any oil-burning furnace or boiler provided that fuel lines and gaskets are not made of natural rubber. Because biodiesel will gel (like No.2 heating oil), storage tanks should be installed indoors or underground. Alternatively, your fuel supplier can provide winter-blended heating oil containing both No.1 and No.2 heating oil with biodiesel blended to B20 to achieve the desired low-temperature storage performance.

Biodiesel Considerations

Before you start filling up the furnace fuel tank, take a few moments to review some of the important issues relating to biodiesel B20 as a heating source:

Biodiesel Properties for Heating Fuel

- contains less than 15 ppm sulphur (an ultra-low sulphur fuel)
- 100% biodegradable
- less toxic than table salt
- high flash point—over 260°F (121°C)
- energy content comparable to No.2 diesel fuel
- reduces furnace, heater, or boiler nozzle cleaning
- works in almost any oil-fired appliance that does not have natural rubber seals and pipes

Disadvantages of Biodiesel

- poor availability
- more expensive than regular petro-heating oil
- gels in very cold weather (similar to No.2 heating oil) and must be blended with No.1 and No. 2 heating oil or stored indoors or underground (Large suppliers will provide the biodiesel pre-blended to suit your climatic conditions.)
- acts as a solvent, removing "crud and sludge" inside fuel storage tanks and lines. Keep spare fuel filters handy and change them a couple of times when first making the conversion to biodiesel blended fuel.

Brew Your Own

People are making their own beer and wine, so why not biodiesel? Many of the diehard hippies and back-to-the-landers have been producing a "biofuel" similar to biodiesel for years. There is a certain charm in going to the local chip truck or greasy spoon to collect their waste cooking grease and turn it into heating and transportation fuel.

Before you consider cutting up the gas credit card and switching to home-brewed "bio" you must consider the cost of processing equipment and materials and the time required to produce the fuel. The transformation of waste oils and greases into biofuel is a relatively easy affair which does not require vast amounts of processing equipment or the budget of Shell Oil. Quality control is another concern, ensuring that batch-to-batch processing is consistent and producing a fuel that is as close to biodiesel as possible. In order for a biofuel to be called biodiesel, certain test criteria have to be met in

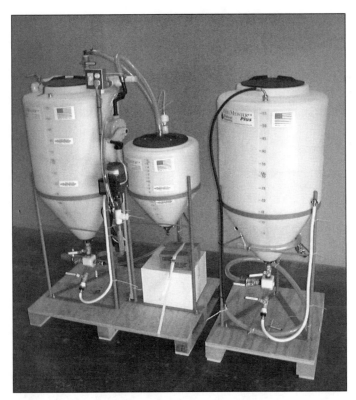

Figure 5-18. If you are so inclined, a biodiesel processor such as this model from Biodiesel Solutions allows you to turn free, waste cooking oil into biofuel. One batch cycle will produce over 40 gallons (151 liters) of fuel with only 24 hours of processing and 1 hour of operator time. (Courtesy Biodiesel Solutions)

accordance with the ASTM test standard D6751 for B100 biodiesel fuels. (See www.astm.org for further details.) Homemade biofuel is unlikely to meet this requirement and using any fuel without this certification rating may negate manufacturer warranties.

Having said that, there are thousands of people mixing their own fuels in garages across the United States and Canada. One company, Biodiesel Solutions of Fremont, California, (www.makebiodiesel.com) offers a complete batch processing system and customer phone/email support line to ensure quality production. According to President Rudi Wiedemann anyone can make their own fuel, and with a supply of waste oil from restaurant sources, production costs are approximately $0.70 per gallon ($0.18 per liter)! At those prices it might just be worth considering getting into production.

Summary

Biodiesel fuel will not solve all of North America's transportation and heating supply issues single-handedly. Other energy options are also required. But since biodiesel is a direct replacement or blending agent with regular petrodiesel or heating oil, no infrastructure or societal changes are necessary in order to start using this fuel source. Biodiesel provides a clean, renewable, domestic and economically diverse energy source which will help pave the way to a more sustainable future.

Figure 5-19. Energy prices are rising, domestic supplies are dwindling, and the environment is taking a beating. Perhaps it's time to consider biofuels for your transportation and home heating needs.

6
Energy-Efficient Transportation

North Americans love their cars... as well as their trucks, minivans, and Sport Utility Vehicles (SUVs). With increasing personal wealth and access to cheap credit, urban vehicular density has exploded. Simultaneously, cities have expanded, sprawling across the landscape as a result of our upwardly mobile culture and our desire to live in the suburbs.

Figure 6-1. North Americans love their cars, and apparently they also enjoy traffic jams and long commutes. Civic planners pay only lip service to controlling the urban sprawl which directly contributes to daily scenes such as this.

The results are not pretty. As a direct result of this increased mobility, the population density of North American cities has declined nearly 50% over the last forty years, decimating historic downtown core areas and local family businesses. As the sprawl continues, city infrastructure costs continue to climb unabated, with an ever larger share of budgets swallowed up for road maintenance and expanded sewer and water pipelines. And yet as we move further and further away from each other we are deprived of green space and our sense of community.

Urban sprawl contributes to environmental degradation: dramatically increased fuel consumption, toxic runoff from road surfaces, and increased landfill waste from decrepit vehicles and auto parts.

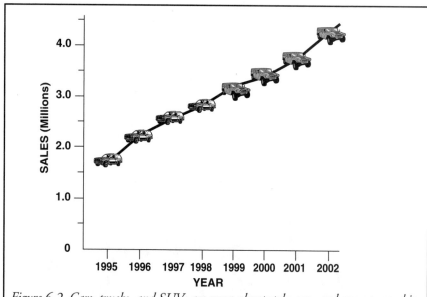

Figure 6-2. Cars, trucks, and SUVs are more about style, ego, and one-upmanship than about transportation. Does anyone really need a 300-horsepower SUV for a city commute? (Data Source: Office of U.S. Transportation Technologies)

Urban sprawl takes its toll in other ways. As the population density of a city decreases, mass transit costs rise exponentially as a result of the need to service a larger area. For example, a city that is 20 miles in diameter (32 km) has an area of 314 mi^2 (804 km^2). If the city grows to a diameter of 40 miles (64 km), the land area increases to a whopping 1257 mi^2 (3217 km^2). This 400% increase in land area necessitates a corresponding increase in the number of buses, rail lines, and other services even though population growth and density do not increase at the same rate.

Urban growth leads directly to an increase in commuting distances, poor quality mass transit (owing to the difficulty of servicing a large area quickly, effectively, and economically), a massive increase in fuel consumption, and many wasted hours spent in cars. Because of the loss of community people don't get to know their neighbours and are reluctant to car pool, resulting in a high percentage of single-occupant vehicles clogging roadways.

There is no doubt that cleaner-emission vehicles, biofuels, and car pooling can help our cities survive the unimaginable increase in the number of personal vehicles on our roads. In the long term, urban mass transit, the development of high density communities, as well as civic will power will be required to set our cities on a sustainable path. Until then, let's consider the *eco-nomic* approach to transportation options:

Fuel Economy and Green House Gas (GHG) Emissions

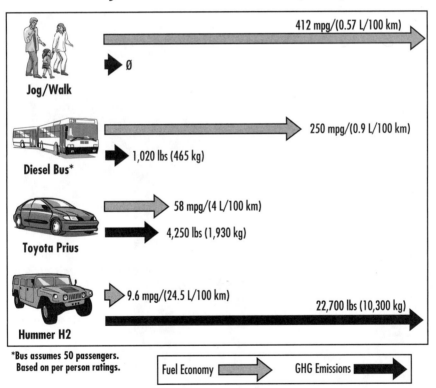

Figure 6-3. This chart compares the fuel consumption and greenhouse gas emissions of four types of urban transportation. Operating costs and greenhouse gas emissions increase in proportion to fuel consumption. Now that celebrity movie stars zip around Hollywood in their hybrid vehicles, driving an SUV to the office is no longer de rigueur.

Bio and Electric Power

Human- and animal-powered transportation has been the norm since time began, and it continues to be the most energy-efficient and clean method of locomotion. Although walking, jogging, and riding a bicycle may not appear to be an option because of urban sprawl, they should not be written off as unrealistic. If we were to convert food calories into gasoline energy, you could bicycle 400 miles (644 km) on the amount of energy contained in a gallon of gasoline. Costs for parking, insurance, and depreciation would be virtually eliminated. So too would greenhouse gas emissions, depending on your penchant for baked beans!

The Pacer scooter shown in Figure 6-6 is the next best thing to walking. Using a 24-volt battery and 300-watt electric motor, the Pacer reaches a top speed of 22 miles per hour (35.4 kph) with an average cruising range of 12 miles (19 km).

Figure 6-4. While this picture might seem laughable, it was only seventy years ago that the depression made this a familiar sight on North American streets. This situation may well return if our appetite for cars and fossil fuel does not abate. A slowly increasing fuel tax spent on mass transit and alternate transportation technologies would help us avoid such a crisis.

Figure 6-5. Think how many acres of land would have to be paved over to park this number of automobiles? Electrically assisted bicycles are the rage in the crowded streets of China. Even if you did own a car, parking it would be as realistic as winning a lottery. (Courtesy Carearth Inc.)

Figure 6-6. (right) Zooming around the city on a folding electric-powered scooter by Pacer is not only cool, it is very energy efficient. This fellow "commutes" all summer long with this model and figures his electricity bill is only a buck or two.

Figure 6-7. (above) This may look like a small motor scooter but it is actually an electrically assisted bicycle. Since it is equipped with pedals the bike can be driven by minors and people who have not bothered with a driver's licence. After a gruelling day at the office, let the electric motor whisk you home. (Courtesy Carearth Inc.)

Figure 6-9. The German made TWIKE is a two-person, three-wheeled, electric/human-powered "car." Although uncommon in North America, Europe has embraced this alternate vehicle for inner-city commuting. It can be powered using an onboard electric motor and integral battery, pedalled, or operated using a combination of the two. Typical range for the 500-lb (227-kg) TWIKE is 25 to 50 miles (40 to 80 km), with a maximum speed of 53 mph (85 kph). (www.twike.ca)

Figure 6-8. No, you don't have to be a circus acrobat to operate the Segway personal transporter. Special "gyroscopic" drive units keep you and the bicycle-like unit upright. Weaving through New York traffic would be a blast on one of these models.

As the vehicles in these pictures illustrate, electrically powered vehicles are not only possible, they are commercially available now. What, then, happened to the mainstream electric automobile? The electrically powered Impulse car manufactured by General Motors in the 1990s looked as if it might just make the grade. It had great styling and good performance, but it couldn't supply the operating range demanded by today's fussy consumer, even though it met 90% of a homeowner's commuting requirements.

Gasoline is a tough fuel to beat. One U.S. gallon contains 129.4 mega joules of energy, which is equivalent to 36 kWh of electricity (34.2 mega joules per liter equalling 9.5 kWh). To put these figures in perspective, the battery bank shown in Figure 6-11 contains approximately the same amount of electrical energy as 1 gallon (3.79 L) of gasoline. The battery bank measures approximately 6' high by 10' long by 2' deep (1.8 m x 3 m x 0.6 m) and weighs in at a whopping 3,800 pounds (1724 kg). In contrast, a gallon of gasoline weighs approximately 7 lbs (3.2 kg).

Figure 6-11. This massive battery bank weighs in at a whopping 3,800 pounds (1,724 kg) and contains approximately the same amount of energy as a gallon of gasoline, which weighs 7 pounds (3.2 kg).

Figure 6-10. Darth Vader meets General Motors - not quite. This solar-powered racing vehicle made by engineering students at Queen's University in Canada is ultra energy efficient, using only sunlight to power it to a top speed of 65 mph (105 kph). Universities around the world develop and build the advanced electronic and mechanical systems and then race their vehicles in numerous world-wide competitions. These races promote cooperation through competition and help to develop technologies that eventually make their way into the mainstream. The solar racers are always a crowd pleaser in whatever cities and towns they pass through.

There is an "underground" economy comprised of small businesses and hobbyists who are converting existing automobiles to electric models. These work well in some applications and have filled the gap created with the demise of the Impulse model. People using forklifts, golf carts, wheelchairs, and other specialty applications would be hard-pressed to find suitable alternatives to the deep-cycle lead-acid battery.

After a hundred years of being addicted to the high energy density and long-range capability of today's road vehicles, consumers simply will not return to the "shortcomings" of electric vehicles. However, considerable research is being conducted on advanced battery technologies, and there is a glimmer of hope on the horizon that suitable range and operating characteristics of electric vehicles may not be far off.

Electrovaya, based in Toronto, Canada, showcased its prototype zero-emission vehicle at the 2004 Better Transportation Expo. Built on a small

SUV chassis, the prototype uses Electrovaya's lithium ion SuperPolymer® battery technology, which offers superior range and performance. This high energy density battery is five times lighter than a typical lead-acid battery and takes considerably less space. According to the company, the price of the all-electric vehicle is still higher than that of standard-production models, but Electrovaya expects prices will drop as production volume rises.

Mass Transit

Properly designed mass transit systems work wonders in high population density urban centers. As the city sprawls into the suburbs, population density drops, reducing the effectiveness of the system and frequently requiring multiple transfers from branch to main-line routes. To circumvent this problem some municipalities have developed "park and ride" facilities which allow rural or suburban clients to park their vehicles (often for free or at a reduced rate) next to a main-line hub. Although this is not the perfect solution, it does provide automotive freedom along with the elimination of both the hassle of driving in crowded rush-hour conditions and the high cost of parking in the downtown core.

Regardless of whether the commute is made by bus, streetcar, subway, or train, the energy-efficiency gains and resulting greenhouse gas emission reductions are achieved by dividing the fuel consumed by the number of passengers on board. Figure 6-3 indicates that a standard diesel-powered bus with an average fuel economy of 5.0 miles per gallon (47 L/100 km) will achieve the equivalent of 250 miles per gallon (0.9 L/100 km) when averaged over a ridership of 50 passengers.

Figure 6-12. Although this bus achieves an average fuel economy of 5 miles per gallon (47 L per 100 km), greenhouse gas emissions are dramatically reduced with the use of biodiesel fuel (see Chapter 5). Passengers on this bus can further increase their fuel efficiency and reduce emissions by taking advantage of the "Rack and Roll" bicycle-carrier system installed on the front of the bus. (Courtesy Kingston Transit)

General Motors has recently developed a hybrid bus technology which bridges the *eco-nomic* gap by increasing fuel economy by up to 60% while keeping capital costs within reason. Buses delivered to the Seattle, Washington area are estimated to save 750,000 gallons (2.8 million L) of diesel fuel per year. According to General Motors, if nine other major metropolitan cities adopted this technology for their buses it could mean savings of more than 40,000,000 gallons (151 million L) of fuel per year.

Figure 6-13. This General Motors hybrid-powered bus delivers up to 60% better fuel economy than conventional diesel systems currently in use. Buses equipped with a hybrid system produce much lower hydrocarbon and carbon monoxide emissions than normal diesel buses, lowering particulate emissions by 90% and nitrogen oxide emissions by up to 50%. (Courtesy General Motors Corporation)

Figure 6-14. Hybrid systems use two sources of power to move the vehicle. Using the parallel hybrid approach the diesel engine acts as a generator, charging a battery bank located on the roof of the bus. Acceleration is achieved using energy stored in the battery bank, with the diesel engine maintaining speed after the vehicle is moving. (Courtesy General Motors Corporation)

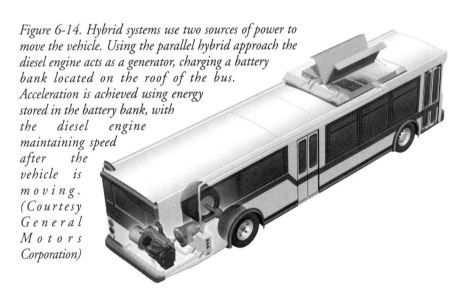

These buses will also improve air quality, producing 90% fewer particulates (tiny pieces of soot and dust which cause lung irritation and asthma) and 60% fewer nitrogen oxides than their standard diesel-powered cousins.

Hybrid systems use two sources of power to move the vehicle. Using the parallel hybrid approach the diesel engine acts as a generator, charging a battery bank located on the roof of the bus. Acceleration is achieved using energy stored in the battery bank, with the diesel engine maintaining speed after the vehicle is moving. A process known as regenerative braking captures energy normally wasted as brake pad heat and returns it to the vehicle's battery bank. In addition to improving fuel economy, regenerative braking reduces wear and extends brake pad life.

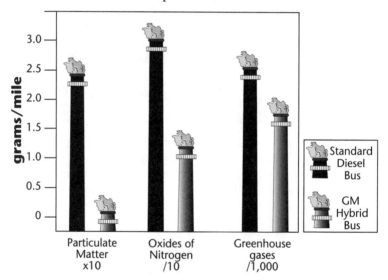

Figure 6-15. Impressive reductions in greenhouse gas emissions, particulate matter, and nitrogen oxides can be achieved with GM's new road-ready hybrid electric bus technology. Fuelling the bus with biodiesel would further improve these gains. (Courtesy General Motors Corporation)

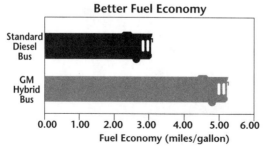

Figure 6-16. According to General Motors, if nine other major metropolitan cities adopted this hybrid technology for their buses it could mean savings of more than 40,000,000 gallons (151 million L) of fuel per year. (Courtesy General Motors Corporation)

Ultra-Low Fuel Vehicles

If your idea of personal transportation is based on protection of the environment and using the most economically efficient means of getting from here to there (assuming that bus, subway, and train are not viable options), ultra-low fuel consumption vehicles might be for you. For years manufacturers have been offering cars with reduced weight, engine displacement, and vehicle size which fall within this automotive class.

The Volkswagen Beetle of the 1960s was one of the pioneers in this field that you either loved or hated depending upon your point of view. The recently resurrected Beetle has been completely redesigned into an automobile that actually works. Coupled with excellent functionality, the gasoline or turbo-diesel versions of this car certainly offer excellent *eco-nomics* and fit the ultra-low fuel vehicle category.

Figure 6-17. The Mercedes Smart™ car is super cute, very safe, and ultra-fuel efficient. The 0.8 litre CDI direct injection diesel engine uses the latest engine technology, greatly reducing noise, smoke, and cold-starting issues. Be prepared to do a lot of waving!

The Smart™ family of automobiles manufactured by the Mercedes-Benz Group combines sporty looks, safety, and economy all in one super cute package. These models use the newly developed CDI (common-rail direct injection) engine that greatly reduces engine noise, smoke, and cold-starting issues. The thrifty little 3-cylinder engine, equipped with turbocharger and air cooler, weighs in at a mere 152 pounds (69 kg), providing fuel economy of 69 miles per gallon (3.4 L/100km). Using such a small amount of fuel means that CO_2 emissions are a negligible 5.1 ounces per mile (90 g/km).

Are you concerned about safety? The Smart™ car features a "steel safety cell" which is reinforced with high-strength steel, making the vehicle one of the safest small cars available according to the manufacturer. Tests conducted by Transport Canada, the country's vehicle safety testing agency, confirm that the car is safe, certifying it for use in Canada. (At the time of writing the Smart™ car is not available in the United States).

Figure 6-18. Sales of ultra-low fuel vehicles have taken off in Europe owing to higher fuel prices. Financially savvy shoppers are beginning to follow suit in North America as a means of reining in spiralling energy costs and protecting the environment. (Courtesy Smart™ Company).

In typical Mercedes fashion the Smart™ car is equipped with numerous features to help prevent road accidents, including antilock braking, acceleration skid control, engine torque control, and passenger airbags. The car is equipped with rear wheel drive with the engine located immediately above the rear axle, offering excellent weight distribution and good performance in wet, snowy, or icy conditions.

With current developments in material and engineering, not to mention superior styling, today's ultra-low fuel vehicles offer a viable alternative for urban commuting. After all, who really needs an SUV to carry their lunch to work?

Hybrid Power Systems

The first question people ask when they hear about gas/electric hybrid-powered cars is how long they have to be plugged-in to charge the batteries. The next question is how far you can drive the car before the battery needs re-

Figure 6-19. If hobnobbing with Hollywood stars is your thing you'll want to scoot right out and purchase a Toyota Prius hybrid automobile. Perhaps then someone will mistake you for Cameron Diaz, Leonardo DiCaprio or Brad Pitt. (Courtesy Ellen and Jerry Horak)

charging. Even though hybrid technology has been around for a number of years and is available in a surprising number of cars, light-duty trucks, and buses, few people understand the principles behind this amazing technology.

The newly updated 2004 Prius is the first midsize hybrid vehicle to en-

Figure 6-20. The 2004 Toyota Prius is the first midsize sedan hybrid vehicle to enter mass production. (Courtesy Toyota Canada Inc.)

Figure 6-21. Being a "full hybrid system" allows the Toyota Prius to operate using either an electric motor, a gas engine, or both power sources simultaneously. This means there is no engine to start, eliminating the need to turn the key. (Courtesy Toyota Canada Inc.)

Figure 6-22. Gone too is the traditional mechanical transmission, which is replaced with a continuously variable ratio drive unit. The Prius is equipped with an electronic joystick instead of a traditional stick shift. (Courtesy Toyota Canada Inc.)

Figure 6-23. All hybrid-powered vehicles are more about electronics and software than brute force. One side benefit of all this electronic wizardry is an information display and touch screen which provides up-to-the-minute fuel and efficiency data as well as providing control for the air conditioning and stereo systems. (Courtesy Toyota Canada Inc.)

Figure 6-24. Other manufacturers are also in the hybrid game and industry forecasts speculate that most automobiles will be available in a hybrid configuration within the next decade. The Honda Civic (above) and Insight (right) models have been available for several years. (Courtesy Honda Motor Co.)

ter mass production. This model employs the hybrid "synergy drive," Toyota's third-generation gas/electric power train technology which improves on earlier generations released in 1997 and 2000. The synergy-drive power train produces more power than previous models, giving the Prius performance comparable to that of similar four-cylinder midsize automobiles while offering an impressive 58 mpg (4 L/100 km) fuel consumption.

The Nuts and Bolts of Hybrid Technology

Automotive hybrid power systems are available in three major design configurations. These are parallel, series, and series/parallel systems, each of which utilizes the same building blocks of a small fossil-fuel engine, an electric motor/generator, an inverter control unit, and a battery bank.

Parallel hybrid configurations connect both the fossil-fuel engine and the electric motor to the drive wheels, allowing energy from either or both sources to move the vehicle depending on driving conditions. While coasting or decelerating, the electric drive motor is actually being driven, and thereby acts as a generator which charges the batteries. The parallel system is relatively simple but has the disadvantage of not being able to simultaneously charge the battery and electrically power the vehicle.

Series hybrid systems are configured so that the fossil-fuel engine drives

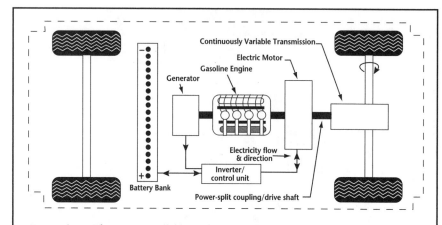

Figure 6-25. The series/parallel hybrid power system improves automotive operating efficiency, reducing fuel consumption through computerized control of an electric motor which is coupled to a small fossil fuel-powered engine. Using brains rather than brawn improves fuel consumption by 50% and reduces greenhouse gas and atmospheric pollutants by almost 90%.

an electric generator, charging the battery bank and simultaneously powering an electric motor to drive the wheels of the vehicle. The advantage of the series system is that a relatively small fossil-fuel engine can operate at its maximum efficiency level, simultaneously powering the car and charging the battery bank.

The series/parallel integrates the features of both the series hybrid and parallel hybrid designs, maximizing system efficiency. An example of a series/parallel hybrid system is shown in Figure 6-25.

In a series/parallel system, a small gasoline or diesel engine is mechanically arranged to drive an electrical generator and provide input to the vehicle's transmission (which provides output power to the drive wheels). An electric motor is mechanically coupled to the input of the vehicle's transmission through a "power sharing coupling."

The battery bank is connected to an inverter control unit which in turn is coupled to the generator and electric drive motor. The inverter control unit is configured to direct the flow of energy through the system depending on driving and operating conditions. For example, when the car is first "started" electricity from the battery bank will flow to the inverter control unit which in turn powers the car's dashboard, driving lights, and air conditioning system.

If the driver *gently* presses the accelerator pedal there may be sufficient energy stored in the battery bank to power the electric motor directly, accelerating the vehicle to driving speed. In this instance the gasoline engine remains off. On the other hand, if the driver wishes to accelerate rapidly, the electric motor will be driven in the same manner as above, but the gasoline engine will start instantly and provide power to assist the electric motor in moving the vehicle.

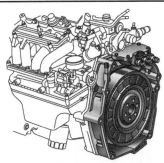

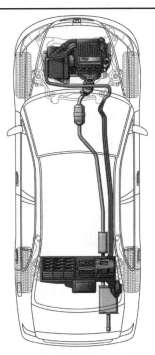

Figure 6-26. The Honda gasoline engine (above) is mated to an electric motor/generator which can provide additional muscle on demand. The phantom view of the Honda Civic hybrid details the layout of the various components. The gasoline engine with its electric motor is mounted above the front wheels. A power cable connects the electric motor to the battery bank and control unit located below the rear deck of the back seats. (Courtesy Honda Motor Co.)

Energy that is used to accelerate a vehicle must be dissipated in order to slow it down or bring it to a stop. A typical automobile will dissipate this energy as heat in the brake pad and rotor assemblies. During braking the series/parallel system uses the inertia of the moving vehicle to reverse the flow of energy through the transmission, in turn operating the electric generator and charging the battery bank.

Large Hybrid Vehicles

General Motors' near-term development program revolves around improving today's automotive technologies. These innovations include electric power steering, more efficient alternators, clean diesel engines, and a variable displacement engine control system known as Displacement on Demand. The company is also developing three different hybrid systems on several of its most popular models of cars as well as light- and heavy-duty trucks.

In 2003 a flywheel alternator/starter hybrid system went into production for fleet customers with retail sales in 2004. The truck's hybrid system links a 5.3-liter engine with a compact 19 hp (14 kW) electric motor and battery bank. A regenerative braking system recovers energy that would otherwise be lost during braking. When the truck is idling, the engine shuts off to save fuel, while the battery bank supplies power to a 120-volt inverter which may be used to operate power tools and other accessories. Based on third-party testing, fuel efficiency is 15% better than that of standard models.

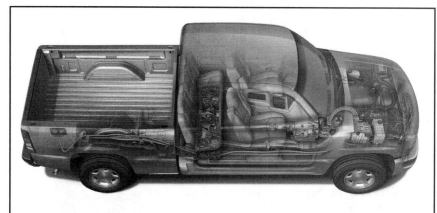

Figure 6-27. General Motors' philosophy is to offer advanced "near-term" hybrid and other technologies to create "no-compromise" vehicles designed for North American driving habits. The flywheel alternator-starter hybrid system introduced on GMC Sierra and Chevy Silverado trucks in 2003 is one of the first products of this philosophy. Based on third-party testing, fuel efficiency for these vehicles has improved by 15%. (Courtesy General Motors Corp.)

General Motors' 2006 Malibu and Saturn VUE models will combine a belt alternator-starter system with their Ecotec four-cylinder engine. Under idling conditions, the engine shuts down and the battery bank powers accessories and air conditioning equipment, providing a 12% increase in fuel economy.

Using expertise developed by GM's Allison division for hybrid buses (described above), full-size sport utilities using this technology will soon become a reality. Beginning in 2007 the Chevrolet Tahoe and GMC Yukon will achieve fuel consumption savings of up to 35% compared with today's models.

Driving a Hybrid Car

Possibly the most disconcerting feature of any hybrid car is the silence. After spending a lifetime in a fossil fuel-powered car, hearing the gasoline engine stop at just about every traffic light brings about a sense of panic and an inclination to push the car off the road. Fortunately this reaction dissipates after the first few hours of driving.

The next most remarkable feature is what you don't feel. Considering the vast amount of computational power and dozens of new technologies making this car operate, everything works normally. The seamless integration of software and machine blends the power from the gasoline engine and electric motor; the driver might just as well be in any typical mid-size sedan.

In an effort to further optimize efficiency, the industry standard four-speed automatic transmission has been replaced with an electronically controlled, infinitely variable model. This arrangement allows the gasoline engine and electric drive motor to operate in their maximum-efficiency zone regardless of road/wheel speed. Additionally, constant velocity transmissions have greater overall energy efficiency than traditional "step shift" models. The observant driver may feel a very slight wobbling effect known as "motor boating" as the transmission continuously adjusts its operating ratios.

Of course the best part of driving a hybrid vehicle is passing the gas station without having to fill up as often. Saving money, reducing atmospheric pollution, and driving Cameron Diaz' car of choice is driving at its best.

Purchasing a Fuel-Efficient Vehicle

Before purchasing any vehicle, determine what your needs really are. Is your search driven by your ego or is your decision based on economics? All road vehicles are expensive, requiring ongoing maintenance, insurance, and fuel in addition to the initial capital purchase. If you are like most people, automotive costs are not tax deductible, requiring the purchase to be made with after-tax dollars. In addition, it is likely that some form of financing will be involved, further adding to the drain on your pocketbook. Consider that you must earn in the neighbourhood of $70,000 before tax to permit you to drive that $40,000 sedan home from the dealer's lot, after taking into account interest and sales taxes. Tack on fuel, insurance, and ongoing maintenance and you may not want to know how much of your discretionary budget is actually spent squiring around town in your Land Rover.

In order to help make rational personal transportation decisions, both the U.S. and Canadian governments provide web sites that guide you through

various questions about automotive ownership, including:
- Do I really need a vehicle or is mass transit a suitable option?
- How large a vehicle do I really need?
- Is it better to purchase new or used?
- What size of engine do I really need?

Once your vehicle needs (and wants) have been determined, both web sites allow you to compare features and fuel economy as well as yearly operating expenses among models. U.S. data is available at www.fueleconomy.gov; for Canadian information visit http://oee1.nrcan.gc.ca.

Car Sharing

Automobile ownership is one of the largest strains on personal discretionary budgets. Most people believe that they *must* own a car even if it is only used to get groceries or go to the theatre a few times a year. If you ask people why they don't simply take a cab, the automatic response is that it is too expensive. Not many people have ever bothered to sit down and do the mathematics of car ownership. As discussed above, purchasing a $40,000 car requires approximately $70,000 in pre-tax income, including financing interest and sales taxes. Fuel, maintenance, and parking add a considerable amount to this figure. Even if you drive an old clunker, operating and maintenance costs may still add up to thousands of dollars per year, and at some point the vehicle will need to be replaced with a newer clunker, requiring a capital investment. By way of example, the Canadian Automobile Association estimates that the annual cost of owning and operating a compact four-door Toyota Echo or similar-size vehicle averages approximately $9,000 per year.

A new phenomenon known as car sharing started in Europe in the late 1980s. For those who have honestly calculated the cost of car ownership and want to cut back on vehicle expenses, this program may be the perfect answer.

You are ready to consider car sharing when:
- you want to reduce driving;
- you want to save money on local transportation;
- you want to avoid the hassle of ownership, including purchasing, repairing, insuring, parking, and other expenses;
- you want to help the environment and improve local community life.

Car sharing is not exactly the same as renting. Members of car-sharing programs usually pay a one-time membership fee and a small access fee based on the estimated number of trips per week. An hourly rate and a mileage fee are then charged based on actual use. For example, VRTUCAR (www.vrtucar.com) charges the frequent driver a monthly access fee of $30, an hourly rate of $2.20, and mileage rate of $0.35 per mile ($0.22 per km). There is no charge for gasoline or insurance.

Cars are strategically located throughout the city and can be reserved on a twenty-four hour a day, seven day a week basis. You can then walk, cycle, or take a bus to the parking area, use a vehicle, return it, and fill in a trip log. A monthly itemized invoice is then sent to your residence.

The VRTUCAR program provides other member services such as reduced rates for mass transit passes, corporate rate structures for longer trips, larger vehicles, and reciprocal membership in car sharing programs in other cities.

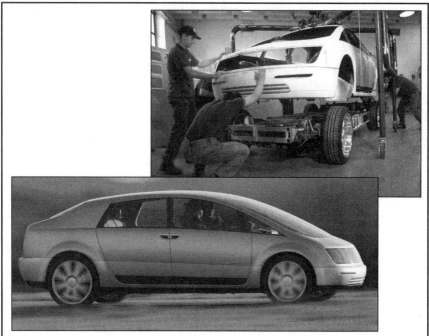

Figure 6-28. The Hy-wire hydrogen/fuel cell car provides a long-term view of the future of automotive technology. The entire propulsion system for this car is housed in an 11" (28 cm) thick skateboard-like chassis. All mechanical linkages for the operation of brakes, transmission, and steering are replaced by software and wires. The engine, steering columns, and other components found in a conventional vehicle are eliminated, allowing unprecedented design freedom and the ability to use one mechanical platform and interchange body styles at will. (Courtesy General Motors Corp.)

The Future of Automobiles

You would think based on TV and magazine ads that hydrogen-powered vehicles were already plying the nation's roads. Sleek cars noiselessly glide down the road while water drips benignly from their tailpipes. Cars that run on water? This sounds more like science fiction than serious automotive technology.

The future of energy no doubt belongs to hydrogen or some other as yet unimagined energy source. However, notwithstanding all of the posturing and hype in the media, the hydrogen infrastructure necessary to power personal vehicles and homes is a long, long way off.

Many people think of hydrogen as a fuel source; it is not. Gasoline, wood, coal, and whale blubber are examples of fuel sources that may be

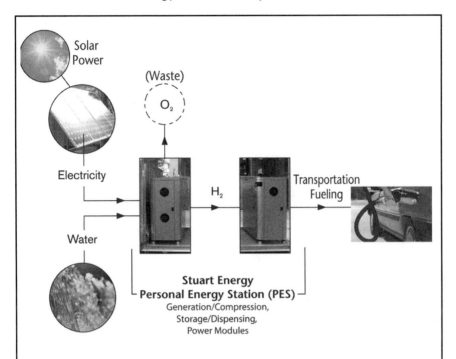

Figure 6-29. Using clean energy from the sun or wind and combining it with water results in clean or "green hydrogen." In theory, hydrogen could be used to fuel our entire economy, heating homes and operating vehicles. Long-term visionaries such as Jeremy Rifkin believe that the nation's automobiles could use their hydrogen fuel to generate electricity for sale to the utility company while sitting parked in the garage. (Courtesy Stuart Energy Systems Corporation)

burned to extract the energy trapped within. Hydrogen, on the other hand, does not freely exist in nature and must be extracted from other sources. In today's economy, approximately 99% of the hydrogen in use is extracted from natural gas, a process which is inherently dirty and provides less energy output than if the natural gas had been burned directly.

Futurists are pinning their hopes on water. Anyone who achieved a passing grade in chemistry will recall the world's most famous chemical symbol: H_2O—plain, clean water. The majority of the earth's surface is covered in water, which also happens to be the largest supply of hydrogen known to man. Ironically, we have been unable to find an economic means of extracting hydrogen from its watery vault.

The most promising means of releasing hydrogen from water is a process known as electrolysis, which on the surface appears to be very simple. Passing an electric current through water causes hydrogen gas and oxygen to be liberated. It stands to reason that if the supply of electricity is inherently clean then the extraction of hydrogen will also be clean, leading to the moniker "green hydrogen," while its natural gas-derived cousin is labelled dirty or "brown hydrogen."

All of this is true provided that the experiments and testing are carried out on a laboratory scale. However, the physics pertaining to the extraction of hydrogen, its molecular size, energy transformation losses, and supply infrastructure problems give us reason to pause. The problems are much deeper than simple economics and cannot be corrected by mass production and the passage of time.

You cannot create something from nothing, and hydrogen extraction is no exception. In order to extract hydrogen from fossil fuels, an inherently dirty and short-term supply approach at best, a given volume of fuel must be "burned" or reformed, producing a volume of hydrogen with lower energy

Figure 6-30. Most of the major automotive manufacturers are cozying up to the long-term view of the hydrogen economy. The Toyota FCHV (Fuel Cell Hybrid Vehicle) proves that the technology works, but the economics is still unknown. (Courtesy Stuart Energy Systems Corporation)

Figure 6-31. Billions of dollars are being spent in research and development covering all aspects of the coming hydrogen economy. Considerable research is required to develop an economic method of extracting clean hydrogen, create a safe storage and delivery infrastructure, and improve the performance and cost of fuel cell technologies. (Courtesy Stuart Energy Systems Corporation)

Fuel efficiency (%) x Vehicle efficiency (%) = Overall efficiency (%)

Overall efficiency

	Fuel efficiency (well-to-tank) (%)	Vehicle efficiency (tank-to-wheel) (%)	Overall efficiency (%) (well-to-wheel)			
			0 10 20 30 40			
Recent gasoline car	88	16	14%			
Prius (before improvement)		28	25%			
Prius (after improvement)	88	32	28%			
Prius with THS II		37	32%			
Toyota FCHV	58 Natural gas-H2	50	29%			
FCHV (target)	70	60	42%			

Note: The Japanese-market Prius was upgraded in August 2002

Figure 6-32. The environmental fuel efficiency of a vehicle must be calculated by analyzing the overall efficiency from oil well to tailpipe. For example, an electrically powered vehicle cannot be considered to have zero emissions if the electricity used to power the car is generated with fossil fuels or coal. This chart, produced by the Toyota Motor Co., indicates that the overall efficiency of a late-model gasoline-powered car is in the order of 14%. In contrast, a 2004 Toyota Prius achieves an overall efficiency of 32%, while a fuel cell-based automobile powered with natural gas-derived hydrogen achieves less efficiency. If "green hydrogen" derived from wind, hydro, or solar technologies becomes a reality it may be possible to reach Toyota's target efficiency of 42%.

content than the original fuel source. Although future consumption of the hydrogen will be clean, greenhouse gas emissions resulting from its production will be higher than if the natural gas had been consumed directly.

Using electricity to electrolyse water drastically reduces or even eliminates greenhouse gas emissions provided that the electricity production is generated by clean energy sources such as wind turbines or hydroelectric stations. Unfortunately, for every kilowatt hour of electricity generated by the clean energy source considerably less than 1 kWh of energy is derived in the form of hydrogen. At the time of writing, wind turbine projects require approximately $0.10 per kilowatt hour revenue in order to be economically viable. Due to production inefficiencies, cost per kWh of hydrogen will move the $0.10 per kWh for wind power considerably higher, making this method of production uneconomical.

Hydrogen gas is the smallest and lightest molecule. Consider this: if a molecule of hydrogen were the size of a basketball, a molecule of gasoline or natural gas would not fit into an SUV when accorded the same scaling factor. Pipes, seals, o-rings, gaskets, tanks, and fittings must contain hydrogen under unimaginably high pressures. Microscopic slits and abrasions in the sealing components seem impenetrable given the massive size of a natural gas molecule, but where hydrogen is concerned they would be like leaving the barn door open.

Stories abound about liquefied hydrogen storage and exotic metal oxide materials that will safely hold this elusive gas. Again, these are nothing but laboratory pipedreams that are no closer to becoming commercially viable than putting a man on Mars.

(I for one believe that a combination of biodiesel (B100) or high-content ethanol (E85) combined with series/parallel hybrid technology will exceed hydrogen *eco-nomics* hands down.)

All of this is not to say that a hydrogen economy will not happen; it is only to say that it will not happen quickly. If you are holding out waiting for the latest in high-efficiency automotive technologies, you need wait no longer. They are here right now—and you don't need to go broke buying in.

7
A Primer for Grid-Connected Electrical Systems

The preceding chapters have focused on ways of saving energy and harvesting thermal energy in the form of green heat and biofuels. It is now time to take a look at renewable energy electrical systems, focusing on electrical generation and how to sell this energy to the utility company.

Generating electricity and selling it to your electrical utility requires an understanding of two separate issues. The first deals with the hardware required to generate, regulate, and feed electricity into the electrical grid. The second issue deals with the regulatory requirements for interconnection and payment for the energy you generate.

Grid-Dependent Systems
Electricity may be sold to the utility using either grid-dependent or grid-interactive configurations. Grid-dependent systems are the least expensive and require virtually no maintenance. An example of such a system is shown in Figure 7-1. In most applications electricity is generated using a photovoltaic (PV) array which captures sunlight and converts it directly to electricity. Although not commonly installed in urban areas, a wind turbine may be used to capture the energy of the wind and convert it into electricity. Wind turbines and PV arrays generate electricity that is not compatible with the utility grid; a device known as an inverter converts this energy into alternating current compatible with the utility system.

Grid Connection PV/Wind Systems
Grid Dependent

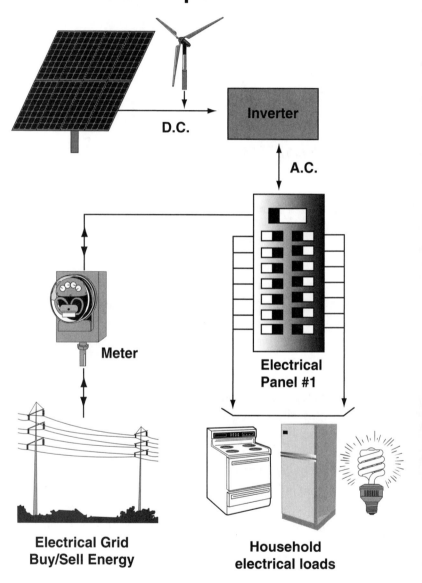

Figure 7-1. The grid-dependent system safely and silently generates electrical power for consumption in the home with any excess production being sold to the electrical utility. This configuration does not provide emergency power to the home during times of electrical blackout. Many jurisdictions in North America recognize the value of this clean-energy generation and offer significant incentives for the installation of grid-dependent systems.

During operation, electricity is fed to household loads first with any excess energy being fed *backwards* through your electrical meter for delivery and sale to the utility. Should the electrical grid fail, safety controls within the inverter will stop electrical generation and disconnect the PV array or wind turbine from the utility distribution system. When grid power returns and stabilizes after the outage period, the inverter will automatically reconnect and start generating and delivering energy. The supply of energy is therefore *dependent* on the *grid* being active, giving us the name of the configuration. An important aspect of grid-dependent systems is the fact that when the grid goes down, so does the power to your home.

Grid-Interactive Systems

The grid-interactive system shown in Figure 7-2 is similar to the grid-dependent system discussed above except that it also has a battery storage bank and a second electrical supply panel powering essential house loads. It is possible to power the entire house, although many urbanites use electricity for very heavy loads such as cooking, hot water, clothes dryers, and air conditioning. With few exceptions, these loads are too large to be economically operated continuously from battery backup.

When the electrical grid is operating normally, power is fed to the utility in the same manner as it is in grid-dependent systems. In addition, a small percentage of the electricity generated by the renewable sources or from the grid itself is used to charge a battery bank. When the electrical grid fails, the inverter disconnects itself from the utility distribution system in the same manner as it does in a grid-dependent design. However, rather than entering an idle mode of operation, the inverter will disconnect the second "essential loads" electrical panel from the utility grid and connect it to the inverter's power output. The inverter will draw power from the renewable sources as well as from a battery storage bank, feeding the loads connected to the second panel. Provided that you have the highest-efficiency appliances and lights throughout the house, relatively normal operation of the home can continue during the blackout.

The amount of time the house can operate in "off-grid" mode is determined by a number of factors. These include the total electrical load of the home, the capacity of the battery storage bank, and electrical generation from the renewable sources. In most applications, off-grid mode is designed to operate the home for very short periods of time, typically measured in hours. This timeline may be increased to days or longer by expanding battery storage capacity, increasing renewable source generation, or reducing

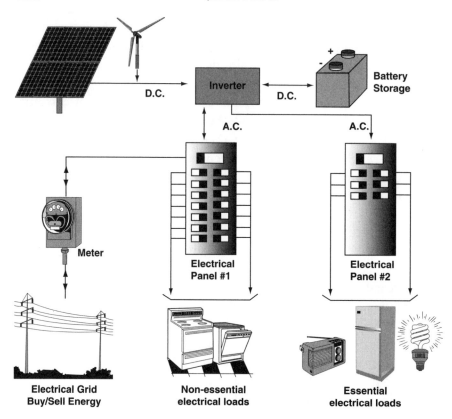

D.C. Inverter D.C. Battery Storage

A.C. A.C.

Meter

Electrical Panel #1

Electrical Panel #2

Electrical Grid
Buy/Sell Energy

Non-essential
electrical loads

Essential
electrical loads

Figure 7-2. Grid-interactive systems work in the same manner as in the grid-dependent configuration in Figure 7-1. The notable exception is their ability to keep the lights on when the grid fails during a power blackout. Electrical energy from the renewable sources and battery storage bank provide power to essential electrical loads in the home.

electrical load. It is also possible to add an emergency backup fossil-fuel generator that can automatically and quickly recharge the battery bank, thereby extending off-grid time indefinitely.

Home and cottage owners who are too far away from the utility distribution grid rely on off-grid systems to provide 100% of their electricity needs. Off-grid systems are also used in recreational vehicles and marine craft where running an extension cord just wouldn't be practical!

Regulatory Requirements

A renewable electricity-generating system will never produce *exactly* the correct amount of power to instantaneously match the electrical loads of the

Figure 7-3. Photovoltaic panels and wind turbines are used extensively in this obvious off-grid installation. What better harmony that using the wind to sail a boat and make electricity at the same time? (Courtesy Southwest Wind Power)

home. For example, a sunny day when everyone is at work and in school will almost certainly generate more electricity than is required by the household electrical loads. Conversely, when the family is cooking dinner and watching television on a dark and rainy night the system will not provide enough electricity to supply these loads.

To solve this problem, many jurisdictions allow the buying and selling of electricity through a program known as net metering. In the example above, excess electricity production is *sold* to the electrical utility by way of the electrical meter spinning backwards.[1] When the renewable sources cannot supply household electrical demand, additional power is purchased from the utility in the normal fashion by making the meter spin forwards.

During the billing period the meter will continue its dance back and forth until it is read by the utility company, at which time it will have recorded the "net" amount of energy purchased during this time period.

Electrical generation beyond what is consumed by the home during the billing period will go into the electrical grid. In some jurisdictions this energy will not be purchased by the utility company. Other utilities will allow "storage" of this energy for a one-year period, at which time they will write a check and zero the customer account. Depending on the utility company, the excess energy may be purchased at the same rate that "normal" utility energy is sold to the customer or at higher "Green Energy" rates, reflecting the value of this clean-energy source.

To further complicate the billing process (but potentially give you more money) *time-of-day* billing exists in some areas. Electricity is treated as a commodity which is purchased and sold through marketing boards just like corn and pork bellies. In other words, the price will rise and fall based on supply and demand. As you can well imagine, the demand for electricity is greatest during the business hours when industry is operating full tilt. Likewise, demand for electricity can drop dramatically during the middle of the night. Electricity prices pretty well match the supply/demand curve.

Renewable energy systems equipped with photovoltaic panels generate their maximum electrical output coincident with daytime peak demand. With a simple net-metering policy, homeowners are paid the same amount for their electricity no matter what time of day it is placed on the grid. Time-of-day billing employs advanced electronic metering to record not only the amount of energy sold to the electrical utility but also when that sale took place.

If you think about this long and hard enough you will notice that most of your electrical consumption will be outside of peak hours, while your maximum generation occurs during peak periods. This creates a form of arbitrage: sell your excess energy at a higher rate during peak hours and buy it back during off-peak periods when the family is at home.

Unfortunately is not possible to generalize about which utility offers what program, or if they even allow the interconnection of renewable-source generation to the electrical grid at all. Before you decide what kind of electrical system you want, contact your electrical utility directly to determine what programs it has in place.[2]

When the Grid Fails

If you have selected a grid-dependent configuration, your home will be in darkness during a power outage just like everyone else's. A look at the discussion in Chapter 13 on fossil-fuel backup generators will help you resolve this dilemma.

---

[1] The example of the meter spinning backwards and forwards is to simplify the concept of buying and selling electricity. In practice, utility companies may require two meters, one for energy sold and the second for energy purchased. Advanced electronic meters may also be used.

[2] An online database for State Incentives for Renewable Energy (DSIRE) is available at www.dsireusa.org. There is no such database for Canadian readers.

In the case of grid-interactive systems, when your home is operating on electrical energy supplied from the grid it is your conscience and your wallet that decide how much energy you consume. Things change quickly when the grid goes down and your home switches to "off-grid" mode of operation. The chips are down at this point and energy efficiency is going to make or break your system.

Energy efficiency and conservation, whether you are on-grid or operating in "off-grid mode" during times of electrical blackout, do not mean you have to live a spartan lifestyle. A big-screen TV, computer, fridge, and central vacuum are examples of appliances that may be found in almost any off-grid home. The trick is not to do without, but to do more with less.

In earlier chapters we discussed how to stretch your electrical energy dollar by reducing electrical consumption without negatively impacting your lifestyle.

All renewable energy systems work in much the same fashion. Collect energy from a renewable source (sun or wind), convert it to utility-grade electricity, supply the household electrical loads, and sell the excess to the

Figure 7-4. Unlike fossil fuels or nuclear energy, renewable energy sources are just that: renewable. No matter how much sunlight we capture with our photovoltaic panels, there will always be more for everyone else; moreover, we're not dumping today's pollution on tomorrow's children. Photo of Colorado Court courtesy Marvin Rand & Associates & PUGH + SCARPA.

utility company. Grid-interactive systems will store some of this energy in a battery bank to be used during times of blackout. When the grid is down, an inverter converts energy from the battery to alternating current (just like utility power) and feeds it to your unsuspecting appliances. Some grid-interactive systems also contain a backup power source, usually in the form of a natural gas generator, or *genset*. These units supply power to charge the battery bank during periods when the renewable system is not able to support the necessary electrical loads, for example during extended grid interruptions or during particularly poor weather.

Figure 7-5. A grid-interactive system requires a means of storing the electrical energy for consumption during a utility blackout. A sizeable battery bank provides the needed storage.

In theory it seems simple, but just like everything else in life the devil is in the details. Figure 7-6 illustrates the operation of a grid-interactive system in more detail.

All of the earth's energy comes from the sun. How do we harness this solar energy? Sunlight shining through a window or on a solar heating panel creates warmth; sun striking a photovoltaic (PV) panel is converted directly into electricity; the sun's energy causes the winds to blow, which moves the blades of a wind turbine, causing a generator shaft to spin and produce electricity; the sun evaporates water and forms the clouds in the sky from which the water, in the form of raindrops, falls back to earth; the rain falling in the mountains becomes a stream that runs downhill into a micro-hydroelectric generator.

While these energy sources are renewable, they are also variable and intermittent. In order to ensure that electricity is available when we need it, a series of wire cables, fuses, and disconnect switches delivers the energy to a battery storage bank.

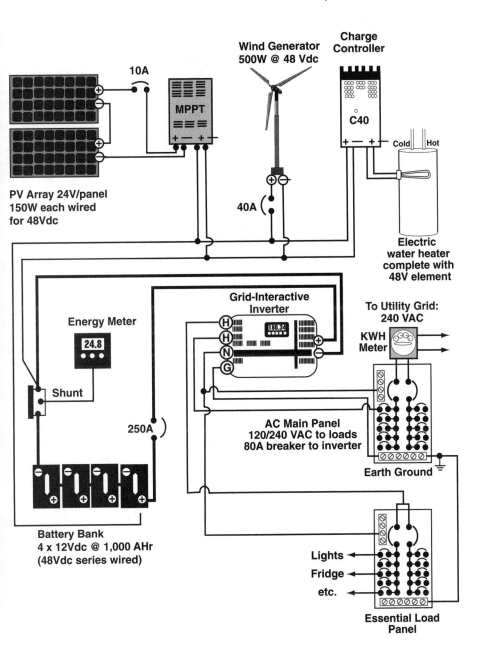

Figure 7-6. Grid-interactive systems provide the additional benefit of operating household loads during blackout periods; however, they require extra planning and have increased system costs as a result of their greater complexity.

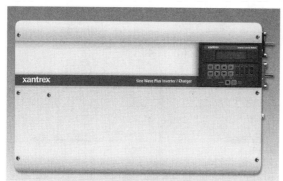

Figure 7-7. Interconnection to the utility grid would be impossible if it weren't for the modern sine wave inverter. (Courtesy Xantrex Technology Inc.)

Although there are many different types of storage batteries in use, the most common and reliable by far is the deep-cycle, lead-acid battery. You may be familiar with smaller ones used in golf carts or warehouse forklift trucks. Batteries allow you to store energy when there is a surplus and hand it out when the grid goes down. So why are we using a battery bank? What other means do we have to store electricity?

Great questions, simple answer: there is not any other feasible method of storing electrical energy. Maybe down the road, but if you want a grid-interactive system now, batteries are the only way to go. Today's industrial deep-cycle batteries are a solid investment that should last twenty years with a minimum amount of care. At the end of its life, an old battery is recycled (giving you back a portion of its value) and a new one is installed. Lower-cost golf cart batteries and sealed "maintenance-free" batteries are also available for smaller systems.

Storing electrical energy is really quite simple: just connect the renewable energy source to the battery and away you go. Getting it out is a bit more complex. First, electricity is stored in a battery at a low voltage or *pressure*. You probably know that most of your household appliances use 120 Volts, whereas batteries commonly store electricity at 12, 24, or 48 Volts. The electricity stored in a battery is in a *direct current* (DC) form. This means that electricity flows *directly* from one terminal to the other terminal of the battery. Direct current loads and batteries are easily identified by a red "+" and black "-"symbol marked near the electrical terminals.

The electricity supplied by the utility to your home is *alternating current* (AC). This means the flow changes direction at a rate of 60 cycles per second or 60 Hertz. (Many of the terms used in electrical generation are named after the early inventors who discovered the physics surrounding a term. James Watt, Count Volta, and Heinrich Hertz are a few of these researchers.)

Figure 7-8. A renewable-energy system should be neat, simple, and well laid out, ensuring smooth sailing with electrical inspectors and your insurance salesman. (Courtesy Outback Power Systems)

In order to increase the voltage (pressure) of the electricity stored in the batteries and convert it from DC to AC, a device known as an *inverter* is required. Without an inverter, selling your renewable energy to the electrical utility would be impossible, as would operating your household appliances during a power blackout.

A supply of electrical energy from the wind or sun feeds low-voltage electricity into a battery bank. The batteries store the electrical energy within the chemistry of the battery "cells." When an electrical load requires energy to operate, current flows from the battery and/or the renewable energy source at low voltage to the inverter. The inverter transforms the DC low voltage to AC higher voltage to feed the "essential loads" house electrical panel and the waiting appliances.

Depending on the complexity of the system, additional devices such as maximum power point trackers, charge controllers, and diversion loads may be required. (Don't worry too much about the terminology just yet, as all of this technology will be reviewed in future chapters.)

The Changing Seasons

As a bad July sunburn will remind you, the amount of sunlight received in summer is much greater than in winter. Simply put, the longer the sun's rays hit a PV panel, the more electricity the panel will push into the utility grid and/or the battery. The months of November and December tend to be dark and dreary by contrast. How does this affect the system operation?

Seasonal variability is extreme in the northeastern section of North America. The two maps in Appendix 5 and Appendix 6 respectively show

Solar Insolation Map
Average Hours of Sun for the Worst Month Yearly

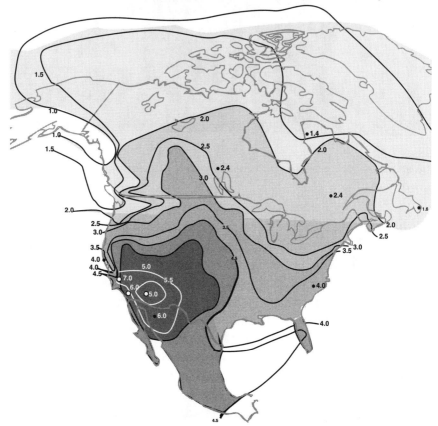

Figure 7-9. Sunlight hours per day are subject to extreme seasonal variability in most of North America.

the average number of sun hours in the worst month and as a yearly average across North America. The amount of sunlight in December is approximately half as much as in September and even less compared to the sunlight in June. Obviously, the output of the PV panel decreases accordingly, and the amount of generated electricity decreases with it.

Although energy production is reduced during the winter months, the interconnection with the grid allows you to purchase any shortfall in household electrical consumption from the utility. In summertime the reverse is true; excess energy production *may* be stored on the grid for consumption in the leaner months. In some jurisdictions storing of excess generation may not be allowed from one billing period to the next, which eliminates the ability to "smooth out" energy production over a yearly period. Be sure to clarify this issue with your utility company prior to purchasing a system.

Back to watts and nuts and bolts for a second. Assume the manufacturers' rating of your PV panel to be 1,200 Watts peak-power output (28 Volts x 43 Amps DC). In reality, PV output tends to be approximately 950 Watts under ideal conditions, less if it is hazy, and nearly zero if the day is cloudy. (Manufacturer rating must be derated for "real world" applications. *Chapter 8 - Photovoltaic Electricity Generation* will explain the reasons why).

Referring to Appendix 5, we find that the worst-month sun hours for a northeastern location averages 2.2 sun hours per day:

2.2 sun hours per day x 950 Watts output = 2,090 Watt-hours per day or approximately 2 kWh per day

During the summer months, the increase in sunlight hours creates considerably more energy:

Production:
6.0 sun hours per day x 950 Watts output = 5,700 Watt-hours per day or approximately 6 kWh per day

You may or may not use all of this energy, depending on household electrical consumption and efficiency. On days when air conditioning and other large electrical loads are idle, the surplus energy is fed to the electrical utility distribution grid via the energy meter.

Phantom Loads

As the name implies, phantom loads are any electrical loads that are not doing immediate work for you, thus wasting energy and dollars. This includes items such as doorbells (did you know that your doorbell is always turned on, drawing power, waiting for someone to push the button?), "instant-on" televisions with remote controls, clock radios, and power adapters.

So what's the big deal? First, these devices are consuming energy without providing much value. That electric toothbrush was probably charged about fifteen minutes after you put it back in the holder. The remaining hours of sitting around until the next time you use it represents wasted energy. A television set that uses a remote control is actually "mostly on" all the time, just waiting to receive the commands from your remote.

While the total dollar cost for these luxuries is small for an on-grid system (around $5.00 to $10.00 per month for an average house), this consumption when operating in off-grid mode during a blackout is unacceptable. By the way, this "small" bit of waste equals millions of dollars and tons of greenhouse gas emissions a month in North America alone.

There is also concern about phantom loads with grid-interactive configurations during a power failure, when they operate in off-grid mode. The inverter unit tries to go into a sleep mode when the last light is turned off at night, conserving a fair bit of energy. Any phantom load keeps the inverter "awake," consuming more energy than it otherwise would. This continuous, unnecessary load on the battery will quickly deplete the bank.

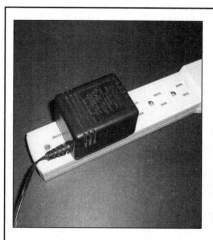

Figure 7-10 A/B. Phantom loads are unnoticed and often useless electrical loads that burn holes in your energy conservation plan. Examples are doorbells, "instant-on" electronics, and microwave ovens or anything that uses a "power adapter," like the one shown above. Locate all of these phantom devices and install them on a power bar or electrical outlet with an on-off switch. Be sure to turn the power bar off when you have finished using any of the appliances.

Phantom Load Management
Some phantom loads can simply be eliminated. Try a doorknocker instead of a doorbell and a battery-powered digital clock instead of the plug-in model. I have even heard that some people use manual toothbrushes.

Television sets and CD, DVD, and VHS players with instant-on and remote control functions should be wired to outlets that can be switched off. This could mean having an electrician wire in a switch for you or using a power bar with an integral on/off switch.

Don't think this is a trivial amount of energy that we are talking about saving. If you add up all of your phantom load-equipped toys, it is possible to burn up nearly a hundred watts of power. Over the course of twenty-four hours, this is a lot of juice:

100 Watts x 24 hours per day = 2,400 Watt-hours or 2.4 kWh/day

A load of 2.4 kWh per day equates to $8.64 per month ($104 per year) of wasted energy, assuming a *delivered* electricity charge of $.12 per kilowatt hour. (The delivered cost of energy is higher than its generation cost due to taxes, transmission, distribution and other surcharges). Think of how many nuclear and coal-powered generating stations, multiplied by the number of households in North America are required just to support this useless load.

Economics

Environmental issues aside, connecting a renewable energy system to the grid is not always economical. What makes system installation economical is net metering and subsidized paybacks in the form of capital equipment rebates. We discussed net metering above. This program allows you to sell power back to the utility at the same retail price you pay. Net metering and "time-of-day" laws are springing up all over the planet and for good reason. Electrical power from centralized fossil fuel plants is an albatross. Power brownouts in California, nuclear generation cost overruns, coal pollution, grid disruption along the northeast coast, and damaging ice storms are all reasons why governments are starting to encourage distributed electrical generation.

At the time of writing, California has the best support program to encourage renewable energy technologies in distributed generation applications. Called "Emerging Renewables Program," it provides a cash rebate of up to 50% of the cost of these systems. For example, a typical grid-dependent system with 2,500 Watts of PV costs the homeowner $7,500.00. The return on investment for this installation can be in the range of 13% to 15% per annum. Where can you safely place your money for that kind of return and increase the value of your home at the same time[3]?

Not all jurisdictions are playing the game, but with energy policies, rates, and fuel costs in turmoil it won't be long before most houses are built with these systems installed. A press release issued by Environment California in July 2004 indicates that officials there are proposing that half of all new homes in the state be running on solar energy within ten years.

Now there's an idea whose time has come.

---

[3] Return on investment (ROI) will depend on energy production per year (sunlight hours per year in the case of PV systems), rate paid by the utility for energy generated, inflation rate for energy and installed capital cost for the system. Discuss these issues with your system dealer and utility to verify you will receive an ROI that meets your investment requirements.

8
Photovoltaic Electricity Generation

The process of capturing and using solar energy is as old as time. Like most things in the modern world, the simplicity of capturing the sun's energy has been elevated to new technological heights. Unlike solar heating collectors that are used to warm fluids running within the collector, photovoltaic cells *magically* convert light from the sun directly into electricity. The photovoltaic cell used in renewable-energy systems is definitely the product of rocket scientists, powering communications satellites to ensure everyone on the planet can watch reruns of *Friends*.

Figure 8-1. Photovoltaic cells are the backbone of renewable-energy systems for homeowners. Since prices and supporting technologies have improved, hundreds of thousands of homeowners are now living lightly on the planet.

The term photovoltaic is derived from the Greek language "photo," meaning light, and "voltaic," voltage which assists the flow of electricity. Friends simply call them "PV" cells for short. Bell Laboratories discovered the PV cell effect in the 1950s. It didn't take the folks at NASA very long to figure out that PV cells would be an ideal means of producing electricity in space. Many missions later, PV cells have improved in performance and have come back down to earth in price. Nowadays, the technology is used in watches, calculators, street signs, and renewable-energy systems for the urban homeowner.

Figure 8-2. PV technology is everywhere you look. Watches, calculators, and even street signs use this ultrareliable technology to provide electricity wherever the economics makes sense.

What is Watt?
For the home renewable-energy system, PV products are relatively standardized, allowing even a novice to make accurate comparisons between product lines. There are currently only three product technologies that should be seriously considered for home use: single crystalline, polycrystalline, and string ribbon cell. Other cell technologies such as thin film or "PV roof shingles" are an option provided product warranty and manufacturer financial strength to honor the warranty period are acceptable. Discussing the cell technology is also a good starting point for understanding how PV cells work.

Figure 8-3. PV cells are similar to transistors, except they're larger. Individual cells are polished, interconnected, and mounted in a frame, creating a PV module.

PV Cell Construction

PV cells are transistors or integrated circuits on steroids. Most people have seen the latest microcomputer chip used in PCs. It's a silicon wafer about the size of your thumbnail that holds several million transistors and other electronic parts. PV cells start out the same way as chip circuits, but they are kept in the oven until they're much larger, approximately 4" (10 cm) in diameter. The baked silicon rods are sliced into thin wafers which are polished and assembled with interconnecting electrical wires.

Figure 8-4. If too much whiskey is added to the batter, PV cells get a bit wavy. Flexible modules are lightweight, easily transported, and rugged, making them ideal for boats, RVs, and trips to the cottage. They can even be fabricated as roofing shingles if you are concerned about "curb appeal." (Courtesy Spheral Solar Power Inc.)

If we were to take one wafer, expose it to bright sunlight, and connect it to an electrical meter, we would measure 0.6 Volts DC (Vdc) of electrical pressure. A voltage higher than the battery rating is required to "push" electrons and charge the battery. For example, to charge a 12 Vdc battery, at least 15 Vdc is required, plus some additional voltage for electrical losses in the system. For this reason, PV cell manufacturers typically connect 36 cells in series to create an additive voltage. (Maybe you should reconsider reading Chapter 1.3 now). A grouping of PV cells thus arranged and mounted in a frame is known as a module.

36 cells in series x 0.6 Vdc per cell = 21.6 Volts Open Circuit

This voltage appears to be a bit higher than our target voltage of "just over 15 V." An interesting phenomenon of PV cells or other electrical generators is a reduction of voltage when the cell is under load, for example when charging a battery. In addition, heat also causes PV cell voltage to drop. When the voltage of a PV cell or series of cells is measured without a connected load, we call this the "open circuit" voltage. Manufacturers often refer to this as "Voc."

When the module is at its maximum-rated power output, the voltage is less than the open circuit rating. This is known as the "voltage at maximum load" and is typically 17 Vdc for a 12 V unit, double for a 24 V model.

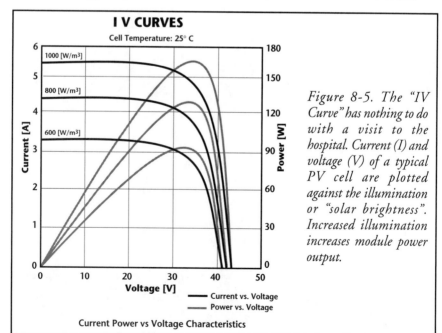

Figure 8-5. The "IV Curve" has nothing to do with a visit to the hospital. Current (I) and voltage (V) of a typical PV cell are plotted against the illumination or "solar brightness". Increased illumination increases module power output.

In order to complete an electric circuit, we must have a source of voltage and current flow. A PV module will cause current to flow within the cells and out of the supply wires to the connected load. As an example, a Sharp (www.sharpusa.com) model NE165 (shown in Figure 8-5) has a rated current of 4.77 Amps (A) at a rated voltage of 34.6 Vdc under ideal conditions and rated illumination.

34.6 Vdc x 4.77 A current flow = 165.04 Watts (W) of output power

This module is therefore rated at 165 W. However, there's one small exception: if it's dark or very cloudy outside, the power output is zero. Obviously light intensity has to play into our equation. The standard light intensity should naturally be that of the sun, as this is the source of energy the panels will use. But where should we measure our sunlight? In Alaska during winter, or perhaps the Bahamas in July? PV manufacturers have taken the liberty of helping us out here. PV lighting intensity has been standardized by industry agreement so that all manufacturers are using the same "solar energy" levels to prepare their data tables. Without this standardization, it would be impossible to determine which PV panel would be the best for your application.

Our sun has very generously decided to output a nice round 1,000 W of energy per square meter at noon on a clear day at sea level. In order to test the intensity of PV panels, light sources have been developed that create this same level of light intensity or *flux*. The light is beamed onto the test PV panel, and its electrical ratings are recorded in the product data sheets. As a consumer, this greatly helps determine the value of one PV panel relative to another.

Figure 8-6. PV systems are upgradeable. This homeowner installed new single-crystalline modules to increase the capacity of the original polycrystalline units located in the center of this array.

PV Output Rating Caution

Use caution when comparing module power ratings. Manufacturer power ratings are based on ideal sunshine conditions, which rarely occur in the real world. It is wise to derate module power ratings by 20-40% based on local atmospheric conditions. Check with local system owners or a reputable dealer to determine the "Real Power Rating" of your proposed installation.

Manufacturer Module Rating x 0.6 = Real Power Rating

...OR...

Real Power Rating ÷ 0.6 = Manufacturer Module Rating

PV module output can also be increased using a technology known as maximum power point tracking. Chapter 11 covers this technology and will discuss how to ensure that your expensive PV panels always operate at their peak efficiency.

PV Module Lifespan

Almost all PV module manufacturers provide a written guarantee for 20-25 years. The manufacturers are obviously quite sure that their products will stand the test of time. The reason for this certainty is the same one that explains why old transistor radios last so long. The semiconductor technology of the cell wafer results in very little wear and degradation.

The standard warranty term from Siemens states that any module that loses more than 10% power output within 10 years or 20% within 25 years will be repaired or replaced. (This is known as a *Limited Liability Warranty*, more commonly known as the fine print; be sure to read this detail carefully to ensure you understand the warranty terms.) Cell technology and quality of workmanship are very high in the industry, so be sure to purchase cells with the best possible warranty for your money.

The cells themselves are quite fragile. To protect them from damage and weather, the cells are bonded to a special tempered-glass surface and sealed using a strong plastic backing material. The entire module is inserted into an aluminum, non-corroding housing to form the finished assembly. Once a grouping of modules, called an array, is mounted to a roof or to a fixed or tracking rack, it should stay put forever.

PV Module Maintenance

Let it rain. That pretty well summarizes what you need to do to maintain your PV array. In the wintertime, ice and snow may build up on the glass. Don't smash the ice to remove it, or you run the risk of smashing the glass too. A quick brushing with a squeegee will take off the loose, highly reflective white snow. Once this coating is removed, the sun will quickly warm the panels, melting any ice even at –4 °F (-20 °C) or lower.

PV Module Installation Checklist

Place your PV module in the sun and collect power—that's it, that's all? Well, not quite. Although PV modules are well designed and last a long time there are several issues that you should considered before choosing where and how to install them:

1. Calculate your electrical-generation requirements.
2. Ensure the site has clear access to the sun.
3. Decide whether to rack or track.
4. Eliminate tree shading.
5. Consider snow and ice buildup in winter.
6. Ensure that modules do not overheat in hot weather.
7. Decide on the system voltage.
8. Locate the array as close to the batteries or inverter as possible.

Step 1 - Calculate Your Energy Requirements

This is square one in your quest for a supply of renewable energy. Let's start the ball rolling with a question: How much energy do you need? If you skipped over the chapter on energy efficiency, it's quite likely that you are consuming significantly more electricity than necessary, resulting in equally high generation requirements.

Start by checking your electrical bills over the last few months or call your utility to find your average monthly electrical consumption. A figure of 20 to 40 kilowatt-hours per day (kWh) is average and may be considerably higher if you have electric heat, electric hot water, or air-conditioning. You should also find out about your utility's policy on grid-interconnected systems. Some utilities will allow you to generate more energy than you consume and "bank" electricity on the grid. At predetermined intervals the utility will reset your meter and send you a check for the excess electricity you have produced. Other jurisdictions may take this excess generation as a free gift, and some utilities are simply ignorant of the whole issue and won't allow interconnection to the grid at all.

These considerations are important variables in determining whether a PV system is what you want and, if so, how large it should be. If your utility pays a reasonable price for excess generation, then by all means produce as much electricity as you can afford. If not, there is no sense in giving away free power. If there is no payment for excess energy, production and consumption of electricity should be as closely matched as possible, with a utility bill of zero being your goal. This is known as an *electrical net zero energy* home.

Figure 8-7. This grid-interconnected, ultra-high-efficiency home is equipped with a 2.4 kW roof-mounted photovoltaic array. It is located in Finland at 63.5 degrees North latitude, proving that photovoltaic panels work in all climates, and not just in the Nevada desert. This home uses one-tenth the heat energy and one-quarter the electrical energy of the typical Finnish home. (Courtesy International Energy Agency)

If your current electrical consumption is a bit on the high side you will quickly find out that you need pockets as deep as Bill Gates in order to generate enough electricity to zero your meter every month. Enter the first rule of energy economics: it is always cheaper to create energy efficiency than to produce additional power. Simply stated, go back to Chapter 2 and determine how you can put your house on an energy diet without affecting your lifestyle.

To quickly recap, let's consider the case for compact fluorescent light bulbs one more time. As discussed earlier, a compact fluorescent light bulb uses approximately 4.5 times less energy than a standard incandescent bulb. Assume for a moment that your home requires 600 W of lighting in the main floor area. If this light is generated using incandescent bulbs, we need 600 W of PV panels to power them. If PV panels are selling for $4.00 per watt, the cost is $2400. With compact fluorescent lamps, however, we need only 130 W of bulbs to create the equivalent brightness, for a PV module cost of $520. Efficiency is **ALWAYS** cheaper than additional generation.

Another factor in determining the size of your PV generation system is any grants or low-interest loans that may be available in your jurisdiction. The California renewables buy-down program provides a 50% grant to the homeowner with a maximum dollar limit. Political realities being what they are, it is always wise to get onto the gravy train while you have the chance.

Lastly, discuss this decision with your dealer. Get advice about system sizes for your roof area, equipment package prices, and other market variables that are impossible to define in this book.

As there is no standard or typical system size, we will assume an electrical energy-generation objective of 4,000 Watt-hours (4 kWh) per day.

Sunlight and PV Energy Generation

The amount of sunlight we receive varies from day to day, depending on clouds, rain, humidity, and smog, but also as a function of the seasons. In most locations, there will be a surplus of sunlight and resulting electrical energy production in the summer, with the opposite being true in the winter. It may be tempting to calculate the energy production of the PV panels using the worst sunlight hours of the year in an attempt to cover all of your energy requirements. Let's take a look at how this would work. Assume you live in the Rochester, New York area. Turn to Appendix 5, *Solar Illumination Map for North America (worst months)*. Find the Rochester area and note that the amount of solar illumination in sunlight hours per day for the worst months of the year is 1.6 hours. We can now plug in some numbers to calculate the size of PV array we require. (Watt-hours per day divided by sun-hours per day calculates the theoretical PV panel size. The theoretical size must then be derated as noted below):

4,000 Watt-hours/day ÷ 1.6 sun hours/day = 2,500 W PV Panel

Don't forget to derate the manufacturer ratings (assume a 20% derating factor):

2,500 real power rating ÷ 0.8 reality factor = 3,125 W manufacturer rating

A quick search of the web for PV panels reveals nothing even closely approaching that figure. The largest 12 V module you can find is from AstroPower and produces around 120 W with a list price of $685.00. Kyocera has an 80 W model on sale for $375.00. This is starting to get complicated and expensive.

PV modules are no different than batteries in the way they can be connected and their capacities increased. Having read Chapter 1.3, you will recall that batteries may be connected in *series* or in *parallel* to increase the voltage or current respectively; the same is true with PV cells. Assuming a PV array with 12 V output, modules can be grouped in parallel to increase the current and wattage rating:

3,125 W Array ÷ 120 W PV module = 26 modules

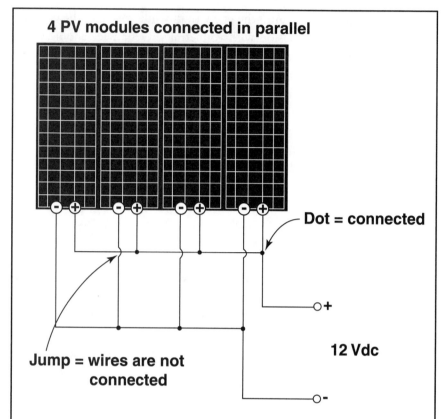

Figure 8-8. PV modules may be connected in parallel to increase wattage in the same way as connecting two small batteries in parallel increases their energy capacity.

Or

3,125 W Array ÷ 80 W PV modules = 39 modules

With so little sunlight in the Rochester area in winter, this PV array will require a lot of modules to support 100% of the electrical load. Now let's take a look at the cost.

26 – AstroPower 120 Watt modules x $685.00 = $17,810.00
39 - Kyocera 80 Watt modules x $375.00 = $14,625.00

Although these figures are inexact because of changing prices, they do reveal an interesting phenomenon. You would expect that since each PV array makes the same amount of power the cost would be the same. Welcome to the world of retail. This is no different than purchasing a car; it's best to shop around.

I will digress a moment to explain how to comparison-shop for PV modules. If you are the sort who cannot understand people who drive a few miles to buy tomatoes at another store because they cost ten cents less, just wait a minute. PV module comparison-shopping is easy and will save you more than enough money to pay for this book and possibly put your kids through college.

The energy-to-cost ratio allows for a very quick comparison of all module types. The calculation is quite simple: divide the cost of the module by the rated wattage. Using the PV modules from the example above:

AstroPower 120 W modules @ $685.00 ÷ 120 W = $5.71 per watt

Kyocera 80 W modules @ $375.00 ÷ 80 W = $4.69 per watt

Depending on the warranty and on the installation cost for the additional modules required to make the same amount of power, the Kyocera modules are probably the least expensive based on the "$/Watt Factor." Now back to Rochester.

It takes quite a few modules to make 100% of the energy required to power a home when the winter sun hours are minimal. The opposite is true in the summer. Refer to Appendix 6, *Solar Illumination Map for North America (yearly average)*. You will find that the number of sun hours per day has nearly doubled. The effect is obvious: when the sun hours per day are peaking, the required number of PV modules in your array will be half the number required at the worst time of the year.

Many system owners and dealers use the monthly averages for one complete year to calculate your energy requirements. Using this approach, generation and consumption of electrical energy will never match on a day-to-day basis. However, over a one-year period your total electrical bill should net to zero. This assumes that your electrical utility's regulations permit "banking" of excess generation during the summer to "fill in" the dips encountered during the dark winter months.

What is the correct number of panels to purchase? There is no "correct" answer. If you have deep pockets, by all means purchase all the modules you require to offset the darkest winter days. If you are like most people and live on a tighter budget, start small and expand your system later as you can afford it. When you win the lottery or your great aunt leaves you some valuable stocks and bonds, splurge and purchase the additional twelve modules that you really need. The additional panels can always be added later, like those shown in Figure 8-6. Just remember to plan for the expansion at the start of the installation. This will lower future costs and headaches.

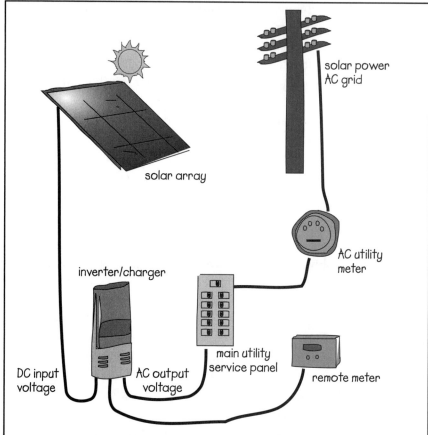

solar power
AC grid

solar array

inverter/charger

AC utility
meter

DC input
voltage

AC output
voltage

main utility
service panel

remote meter

Figure 8-9. A grid-interconnected renewable energy system will have a positive impact on your electrical bill as well as on the environment, regardless of its capacity. Most designs try to balance generation with daily consumption. (Courtesy Xantrex Technology Inc.)

Step 2 – Ensure the Site Has Clear Access to the Sun

PV modules must face the sun to make electricity. As every romantic knows, the sun sets in a brilliant display of color in the west. My mother-in-law tells me that it rises in the east, although I cannot personally attest to this statement. Where, then, do you face the modules? The quick answer is to point the panels directly at the sun. However, with the sun moving from east to west during the day and progressively higher in the sky as summer approaches, this might be a rather difficult task.

You have two options: mount the modules to a fixed location pointing solar south or install a tracking unit that automatically aims the panels directly at the sun. Either way, make sure the location you choose is not impeded by trees, buildings, or other obstructions. Site locations can be difficult to assess due to the seasonal variation in the sun's track, leaves on deciduous trees, and neighboring buildings. To help simplify the assessment of a site, consider purchasing or renting a Solar Pathfinder to eliminate any guesswork. This device works on the well-known principal observed by every schoolchild with access to a magnifying glass. As the sun tracks across the sky, light enters the magnifying and focusing dome of the device (see Figure 8-10). The concentrated sunlight falls on specially sensitized paper burning a copy of its path. Gaps in the marking indicate obstructions that will cause unacceptable shading of the PV panels.

Figure 8-10. The Solar Pathfinder is used to determine magnetic and solar south (using the data in Appendix 4) as well as to record the sun's track across the sky using the specially sensitized paper shown right. (Courtesy Solar Pathfinder)

Step 3 – Decide Whether to Rack or Track

Roof Mounts

Perhaps the simplest mounting system is to attach the panels to a solar-south-facing roof as shown in Figure 8-11. Solar south differs from magnetic south as a result of a phenomenon known as magnetic declination. Appendix 4 shows the correction of compass readings based on your geographic location. If you live in Florida, for example, there is no change between magnetic and solar south. If you live in Alberta, the error in the compass-magnetic versus solar south reading is so large that it will affect your system's energy production if you use fixed-mount PV arrays.

Figure 8-11. Attaching a PV array to a solar-south-facing roof is one of the least expensive mounting methods. The downside can be getting on the roof to brush off a foot of snow in the middle of winter.

Another problem with fixed-mount arrays is their inability to change their angle of view throughout the seasons. The winter sun barely scrapes across the horizon in most of the northern United States and Canada. During the summer, the change in the earth's angle places the sun almost directly overhead. If your roof mount or other fixed mount cannot be adjusted for summer-to-winter sun angles, a good rule of thumb is to set the array at an angle from the ground equal to your latitude. Mounting racks, which have a summer/winter angle adjustment, should be lowered by fifteen degrees for summer. Likewise, raise the angle by fifteen degrees for winter.

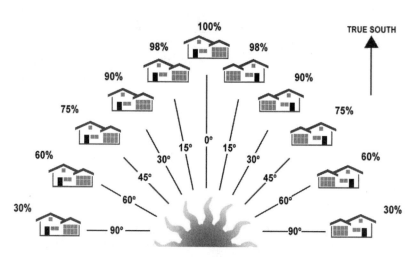

Figure 8-12. PV panels should be mounted facing as close to solar south as possible. PV panels mounted 45° east or west of solar south will reduce potential electrical output by 25%.

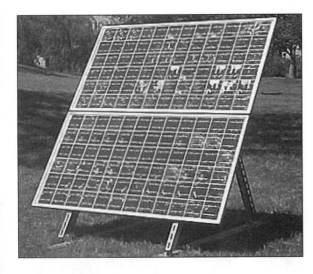

Figure 8-13. The commercial ground-mounting unit is ideal for smaller PV arrays and is also a good option where cost is a concern. (Courtesy Zomeworks)

Ground Mounts

If the thought of climbing onto your roof to wipe snow off the PV array has you going woozy, then ground mounting is for you. A ground-mounted PV array is a simple and inexpensive method of mounting the panels. The unit shown in Figure 8-13 was purchased, but you can also build a unit yourself. Select galvanized steel or aluminum to limit corrosion. When designing your mounting system, make sure to include a hinge assembly to allow for seasonal adjustment. It is possible to make a ground mount using preserved wood or cedar, but keep in mind that the panels will almost certainly outlast the wood.

A word of caution regarding ground mounting: PV modules are expensive and may have a tendency to walk away. Ensure that you have security bolts (or a good guard dog) to prevent theft. If possible, place the mounting legs in concrete footings to make sure that everything stays in its place. Snow is another problem. If you live in an area where "sweeping" the array requires a snow blower rather than a broom, then ground mounting is not for you. Be careful to ensure that lawnmowers and playing children will not send rocks or other debris flying at the modules. Although it is very likely that no damage will result, there is no sense tempting fate.

Tracking Mounts

The tracking mount shown in Figure 8-14 is the most advanced means of pointing your PV panels at the sun and has the added benefit of being hypnotic to watch as it scans the sky. PV arrays comprising up to twenty panels are mounted on the tracker, forming a billboard-sized unit. (A sixteen-module array measures 16' x 8' or 4.9 m x 2.4 m.) The tracker is designed to move to an easterly location when it gets dark. At first light it follows the sun on its westerly track across the sky. At day's end, the unit returns to the easterly heading to repeat the process. Electronically controlled trackers can be fitted with a manual seasonal-adjustment device or with an automatic version.

Figure 8-14. Tracking units such as these increase summer electrical production by up to 50% but offer little improvement during the winter.

Passively controlled trackers are also available. The Track Rack manufactured by Zomeworks (www.zomeworks.com) uses heat from the sun to operate a tracking mechanism. Although this mechanism is simpler than those used in active tracking units, the Track Rack model will go to sleep facing west and may require an hour or two of valuable sunlight before it will start tracking correctly on cold days.

Should you use a tracker? The debate rages on, but the following are considerations that may affect your decision:

- Trackers increase summer PV production by up to 50%. Winter production is improved only 10-20% due to the lower, smaller arc of the sun.
- The further north you are the less sense it makes to track in the wintertime. This is especially true along the United States/Canada border area.
- If your site has a limited window of sunlight—less than six hours— then tracking will not greatly improve system performance.
- Trackers are not cheap. The cost of the tracker may be used to pur-

chase a fixed rack and more PV panels, which might offset the loss in non-tracking production.

- Trackers add a degree of complexity to the system. The bits and pieces are just one more thing to have to maintain.

Regardless of which type of PV mounting system you decide to use, keep in mind that they take up a fair bit of area and make wonderful kites or sails in high winds. Ensure that proper mounting and foundation work has been undertaken in compliance with the manufacturer's installation instructions. If you want to harness the wind, don't use your PV panels.

Step 4 – Eliminate Tree Shading

Although you want to protect the south side of your house from the summer sun, the last thing you want to do is to shade even a very small portion of your PV array. Partial shading will cause a disproportionate reduction in electrical generation to the shaded area. Keep trees well clipped in the sunlight window (between the hours of 9am to 3 pm), keeping in mind both winter and summer sun tracking. If you are unsure about shading, consider evaluating your site using the Solar Pathfinder shown in Figure 8-10a.

Step 5 - Consider Snow and Ice Buildup in Winter

If you live in a snowy area, take winter snow and ice buildup into account when considering PV location. Although PV angles at this time of year are almost vertical, snow and ice will stick to the array. If a coating of fresh, white, powdery snow is covering the array, it may take several days for it to fall off without brushing—and you may not want to brush off PV arrays if they're mounted on a second-story roof.

Step 6- Ensure That Modules Don't Overheat in Hot Weather

PV module electrical output fades (just like everyone I know) as the mercury rises. In order to ensure peak operation of the array, modules must not be seated directly on the roof surface. An air gap of 2-3" (5-8 cm) will allow cooling air to circulate under the array, providing maximum power output.

Step 7 – Decide on System Voltage

For grid-interconnected systems, there are two commonly used voltages: 24 or 48 V. The standard in the PV industry is modules with 12 or 24 V nominal output. As discussed earlier and as shown in Figure 8-8, PV modules can be interconnected in parallel to increase current flow and system wattage without changing the voltage of the panels. Figure 8-15 illustrates a mixture

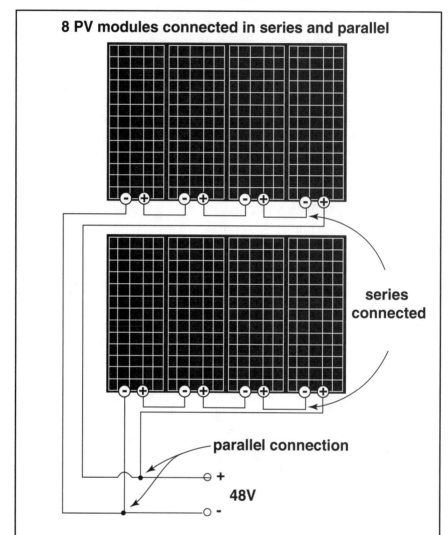

8 PV modules connected in series and parallel

series connected

parallel connection

+

48V

-

Figure 8-15. PV panels are just like batteries: they can be interconnected in series to increase voltage and in parallel to increase the current of the array.

of series and parallel-connected PV panels. In this arrangement, there are two rows of four PV panels. Both rows are interconnected in series, increasing the system voltage. If you look carefully and pretend that each PV module is a battery, you can follow the series connection. The positive of one module connects to the negative of the next, and so on. When each 12 volt panel is connected in a series grouping of four, the voltages are additive, resulting in a 48 volt array.

Once both rows of panels are at the same system voltage, which in this example is 48 V, the array may be connected in parallel. Examine the wiring connection at the point indicated in Figure 8-15. The negative lead from each row is connected, as is the positive lead. Together, the two rows of 48 V arrays will deliver additional current flow (power) to the load.

Step 8 - Locate the Array as Close to the Batteries or Inverter as Possible

Have you ever connected 2 or 3 garden hoses together, turned on the tap, and wondered why the sprinkler was dribbling an anemic spray of water even though the pressure at the house was fine? What you have witnessed is resistance to flow. This phenomenon applies not just to garden hoses but to electrical circuits as well. The resistance to flow of electricity works in the same manner. Low-voltage (pressure) circuits require very large wires to carry the necessary current in the cable. It stands to reason that the further the PV array is from the battery or inverter, the greater the loss of electrical energy will be. Increasing the pressure or voltage of the electricity will help, but distance is a bad word when it comes to low-voltage direct current circuits. Keep the wire runs as short as possible and maximize wire size to prevent unnecessary power loss.

As a side note, this is the reason that the North American electrical grid transmits energy at very high voltages, reaching 750,000 volts in some areas. Even at this high voltage electrical losses through the system can be extreme. The province of Ontario has approximately 15,000 miles (24,000 km) of transmission grid. The electrical losses on this system are equal to the energy output of three large nuclear reactors! Distributed generation using grid-interconnected photovoltaic systems places the consumption and generation of electricity at the same location, thereby eliminating transmission losses. At a couple of billion dollars apiece for nuclear generating plants, not to mention fuel disposal, decommissioning costs, and safety liability, PV panels mounted on the nation's roofs are a bargain.

Conclusion

PV systems are simple and reliable. With electrical rates rising and governments committed to "greening the grid," there is little question that all of North America will be pushing PV in the same manner as California is now. According to Environment California, recent proposals before the state legislature would have 50% of all new homes running on solar energy within the next ten years. Although this goal is lofty and well ahead of any other jurisdiction in North America, we still have a long way to go before we catch the leaders, Japan and Germany.

Figure 8-16. The Conde Nast building in Manhattan is a magnificent example of what the future will look like for urban dwellers. This building, equipped with advanced energy-efficient designs, fuel cell electricity generation, and photovoltaic panels, pumps green electricity into the New York State grid. Over the coming years, commercial buildings, condominiums, and apartments will almost achieve a net zero energy goal. (Courtesy Fox & Fowle Architects / Andrew Gordon Photography)

9
The Urban Wind Turbine
a.k.a. *The Urbine*

There must be something about the sight of a wind turbine and its spinning blades that make men's knees weak. Perhaps the technology has been around long enough that it's in our genes, like working a barbecue when we would never consider cooking in the kitchen.

Whatever the case, mankind has been capturing the wind for eons. Every schoolchild knows of the fabled Dutch windmills of old and the gradual evolution of the modern electricity-generating turbines that are quickly dotting the world's *rural* skyline. The key word here is "rural" given all of the limitations of the urban turbine or "urbine".

Home-Sized Wind Turbines
The market today is alive with many different wind turbine systems. There is also a market for rebuilt and used machines that were in use before rural homes were connected to the growing electrical grid in the early 1900s. The resource guide in Appendix 3 provides a listing of the major manufacturers and rebuilders of new and refurbished equipment.

Before we venture further into this chapter, we need to ask an important question: are wind turbines suited to urban applications? This is perhaps one of the most hotly debated topics around, next to picking which stock will be the next Microsoft in your portfolio. Without a doubt, wind turbines have been installed in some of the most interesting, unusual, and controversial places on the planet; such as the tops of trees or billboards. There is also a segment of society that believes wind turbines are really no different from photovoltaic panels, a technology that can be installed just about anywhere.

In a nutshell, the *urbine* is a mythological beast of the wind not unlike Pegasus. Installing and operating an urban-based wind turbine is not worth the hassle, even if they worked properly. Dollar for dollar and watt for watt, the photovoltaic panel beats the urban wind turbine hands down. The following discussion explains why.

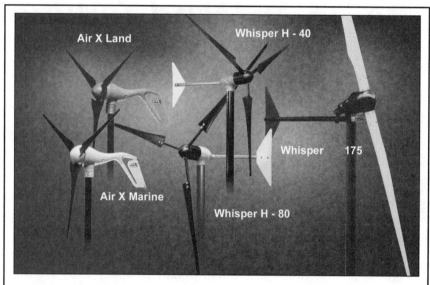

Figure 9-1 Wind turbines are available in a wide variety of sizes, designed to suit just about every application. (Courtesy Southwest Wind Power)

Site Location and Installation

A major difference between photovoltaic and wind-based systems is their size and height. You can hang a few photovoltaic panels on almost any roof, while wind turbines will attract attention for miles around. If those "miles around" happen to be forest or grazing land, then only the cattle will pay attention. But don't even consider installing a wind turbine until you have all the "i's dotted and t's crossed." Not everyone will share your enthusiasm for wind technologies, especially if your neighbor is located just a whisper away.

Where you locate a wind turbine will generally be determined by your lot size, cable distance to your home, and tower construction. The best location, however, is where the wind blows strongest. Wind speed and

Figure 9-2. Wind turbines are used extensively in "off-the-grid" applications such as the rural home shown here. A rural homeowner has better control of the surrounding environment than his urban cousin, ensuring that wind energy is smooth and continuous, without the effects of neighbouring buildings and trees. (Courtesy Southwest Wind Power)

smoothness (or laminar flow) is enhanced over unobstructed grazing land but impeded by neighboring rooftops, owing to ground-surface smoothness. Homes dotting the suburban landscape increase the wind's "working" ground level to the roof height of these homes. The effect of numerous rooftops impeding the wind flow increases turbulence and reduces the available energy delivered to the turbine. Additionally, rough airflow causes the turbine to *work* harder, constantly "yawing" and accelerating. The net effect of all this bucking and twisting is reduced turbine life and energy output.

A general rule of thumb is to mount the turbine on a tower 30' (9 m) above any obstructions that are within a 300' (91 m) radius of the tower. For example, the houses and trees shown in Figure 9-3, which are within this area, must be at least 30' shorter than the overall turbine tower height. Keep in mind that trees grow, so provide extra tower height or use pruning equipment to maintain these minimum dimensions.

The effective tower height (actual tower height minus the working ground level) and laminar airflow influence on turbine power cannot be overemphasized. Increasing tower height from 60' to 120' increases, the power output of the Bergey Excel-S turbine from 330 kWh to 550 kWh per month at 8 mph (13 kph) wind speed. In other words, doubling effective tower height

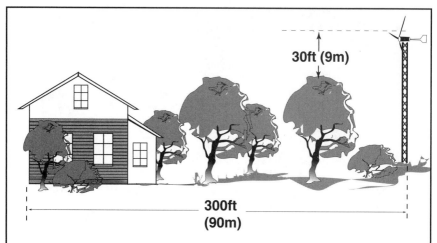

Figure 9-3 A wind turbine must be placed in smooth air, clear of any obstructions, in order to ensure maximum power output and life expectancy.

increases power output by a factor of 1.7 times. Keep in mind, however, that doubling tower height to increase the working ground level increases tower and installation costs and creates an urban eyesore for your grid-connected neighbours.

Noise and Mechanical Loads

Wind turbines differ from photovoltaic panels in another key area. The rotating and yawing mechanical components create vibration and stresses that result in noise and mechanical loads on the mounting structure. If the structure happens to be a well-engineered tower located away from your residence, there is no problem. However, attempting to install a wind turbine on the typical suburban home will generate appreciable noise as well as static and dynamic mechanical loads.

Responses to wind turbine noise is extremely varied. Some people feel that the noise is in harmony with nature, while others insist that Spitfires during wartime were not nearly as audible. Perhaps you appreciate the sound, but will your neighbour?

Mechanical loading on your home's structure is another matter. The standard wood-framed house of today was not designed to act as a wind turbine tower. The Southwest Wind Power Air 403s shown in Figure 9-4 are available with a roof-mounting kit that the manufacturer claims absorbs these stresses. Perhaps, but can you be assured that things won't "go bump in the night" the next time a major storm howls through?

Figure 9-4 The Air 403 is about the largest wind turbine that should be considered for roof mounting. If you need a little "white noise" to sleep at night, this is one heck of a way to get it! (Courtesy Southwest Wind Power)

People keep trying to develop the perfect *urbine*, with numerous tests site at universities and research centers around the world. Paul Gipe, a leading expert on wind turbine technology, cites the case of a wind turbine installed on the roof of a Northwestern University engineering building.. The noise generated by the turbine was surprisingly loud, causing the faculty to terminate its use.

The moral of this story is simple. If you are going to consider the installation of an *urbine*, tread carefully. It is a difficult and uncertain path to follow and one that should be avoided unless you have carefully assessed the risks.

Figure 9-5a and b. Wind turbines and universities seem to go hand in hand. This urbine installation was recently completed at the University of Toronto, Canada. Time will tell if the cost and reliability of this installation match those of the maintenance-free photovoltaic panel. (Courtesy True North Power).

This large wind turbine is located on Lake Ontario in Toronto, Ontario Canada. Since many major cities are located along large bodies of water, a traditionally good place for wind resource, it is most economical for many urban dwellers to leave wind power to the experts and purchase it through a green power program. See Chapter 4 on community power.

10
Battery Selection

Why Use Batteries?

Why use a battery bank in the first place? This is a reasonable question with a simple answer. You will be the laughing stock of the neighborhood if you have a grid-connected renewable-energy electrical system and your lights go out when the grid goes down. To prevent such an embarrassment, it might be wise to consider a source of backup power just in case the utility tries to plunge you into darkness.

As you will learn in Chapter 12, inverters are devices which convert the electrical energy from your photovoltaic panels or wind turbine into utility-grade alternating current. This energy is then supplied to the electrical grid. For safety reasons, when the grid fails the inverter must enter an idle state and disconnect from the electrical grid, preventing a condition known as back feeding. If the power failure occurs at night when your photovoltaic panels are asleep, there will be no electricity to operate your inverter and supply household loads. To circumvent this problem, a battery bank can be installed to store sufficient energy to operate essential household appliances and lights during the blackout.

What about Fuel Cells?

Automobile manufacturers are starting to use fuel cells, so why not install some of these? Fuel cells are wonders of technology and work well in specific applications, but home use isn't one of them. Besides, fuel cells don't store energy; they simply convert it from one form to another.

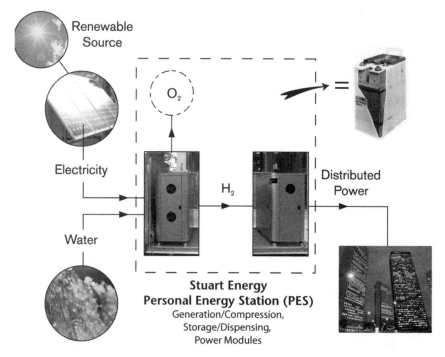

Figure 10-1. Electrical energy from photovoltaic panels combines with water in an electrolyzer device which generates oxygen and hydrogen. The hydrogen must be compressed and stored until it is required. Feeding the hydrogen gas to a fuel cell generates electrical power and waste heat. The poor overall efficiency and the expense of this process mean that fuel cells cannot match the economics of the industrial deep-cycle battery.

Electrical energy from photovoltaic panels and wind turbines combines with water in a device known as an electrolyzer. The electrolyzer generates hydrogen and byproduct gas oxygen. The hydrogen is compressed and stored in a pressurized cylinder until it is required by the fuel cell, where it recombines with atmospheric oxygen, generating electrical power and waste heat. The poor overall efficiency and the expense of this process mean that fuel cells cannot match the economics of today's industrial deep-cycle battery.

Industry fuel cell manufacturer Ballard in conjunction with Coleman produced a small home-sized unit to act as a battery-backup system for computer and office equipment. Unfortunately, safety and supply issues with hydrogen gas as well as capital costs have killed that project for now. If we revisit this in another ten to twenty years the story *might* be a little bit different. However, although fuel cells sound exciting the battery industry is also developing advanced methods of energy storage with lower pricing. The jury is out on which technology will prevail in this particular application.

Battery Selection

There are thousands of batteries available, and the more you look the more selection and complexity you will find. Car, truck, boat, golf cart, and telephone batteries are a few examples that come to mind. Then, of course, there are the "special" kinds such as NiCad, nickel metal hydride, and lithium ion, to name a few. What is the correct choice for our renewable energy system?

First of all, let's eliminate a few of the legends and quick fixes that abound:

- Used batteries of almost any size and type are not the answer. It doesn't matter how good a deal you got, the batteries are cheap because there is a problem with them. All batteries die after a given life span or if incorrectly serviced. Sure, they might work for a year, maybe even two or three, but the lower charge capacity of older cells makes this a sour deal.
- Quality battery costs are high. A golf cart battery of the same capacity as a high-quality "deep-cycle" battery may cost less, but the reduction in life span is not worth the hassle except for very small electrical loads or short grid interruptions. Changing batteries is backbreaking and dangerous work. You only want to do this once or twice in your lifetime, so stick with the best.
- Fancy batteries such as NiCad or lithium ion work well in cell phones and camcorders, but have you ever bought one of these babies? To get a lithium ion battery big enough to operate your renewable-energy system would probably cost as much as your entire house. Be thankful you can have one in your cell phone and leave it at that.

The only type of battery to consider in a full-time renewable-energy system is the deep-cycle lead-acid industrial battery manufactured with liquid or gelled "maintenance-free" electrolyte. These batteries have been around a long time and have been engineered and re-engineered so that they offer the best value for your money. Good-quality liquid electrolyte industrial batteries should last between fifteen and twenty years with a reasonable amount of care. In addition, the batteries are recyclable, and many companies offer trade-in allowances for their worn-out models, which also eliminates an environmental waste issue. The same is not true for NiCad or other exotic blends, which are considered hazardous waste.

How Batteries Work

The typical deep-cycle electrical storage battery, like a car battery, uses a lead-acid composition. A single-cell batter uses a plastic case to hold a grouping of lead plates of slightly different composition. The plates are suspended in

Figure 10-2. These models from Rolls Battery Engineering (www.surrette.com) are typical of long-life renewable-energy-system batteries. The two batteries in the foreground have the same voltage rating (6 Vdc). The model on the right has twice the capacity as the model on the left. (Courtesy Surrette Battery Company)

the case, which is filled with a weak solution of sulfuric acid, called *electrolyte*. The electrolyte may also be manufactured in a gelled form, which prevents spillage. (Batteries of this type are often sold as "maintenance free"). The lead plates are then connected to positive and negative terminals in exactly the same manner as the AA and C cells described in Chapter 1.3. A single-cell with one negative and one positive plate has a nominal rating of 2 V.

Connecting an electrical load to the battery causes sulfur molecules from the electrolyte to bond with the lead plates, releasing electrons. The electrons then flow from the negative terminal through the conductors to the load and back to the positive terminal. This action continues until all of the sulfur molecules are bonded to the lead plate.

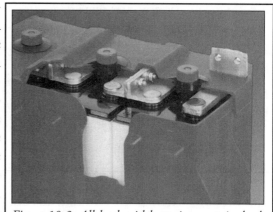

Figure 10-3. All lead-acid batteries comprise lead plates suspended in a weak solution of sulfuric acid. The size of the plate and the acid capacity directly affect the amount of electricity that can be stored. Each cell of the battery can be interconnected with others, increasing capacity and voltage. (Courtesy Surrette Battery Company)

When this occurs, it is said that the cell is discharged or dead.

As the cell is discharged of electrical energy, the acid continues to weaken. Using a device called a hydrometer (see Figure 10-4), we can directly measure the strength or specific gravity of the battery electrolyte. A fully charged battery may have a specific gravity of 1.265 (or 1.265 times the density of pure water). As the battery discharges, the specific gravity continues to drop until the flow of electrons becomes insufficient to operate our loads.

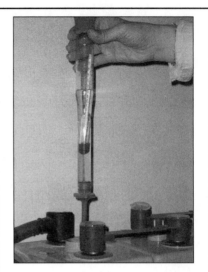

Figure 10-4. The hydrometer measures the density or specific gravity of fluids such as the electrolyte in this cell. The higher the reading, the more electrons are stored in the cell, indicating a higher state of charge.

When a regular AA or C cell is discharged, the process is irreversible, meaning the cell cannot be recharged. Discharging a deep-cycle battery bank is reversible, allowing us to put electrons back into the battery, thus recharging it. Forcing electrons into the battery causes a reversal of the chemical discharge process described above. When a photovoltaic panel is placed in direct sunlight, it generates a voltage. When the voltage or pressure at the PV panel is higher than that of the battery, electrons are forced to flow from the panel into the cell plate. Electrons combine with the sulfur compounds stored on the plate, in turn forcing these compounds back into the electrolyte. This action raises the specific gravity of the sulfuric acid and recharges the battery for future use. Although there are many types of battery chemistry, the charge/discharge concept is similar for all types.

Depth of Discharge
A deep-cycle battery got its name because it is able to withstand severe cycling or draining of the battery. A car battery can only withstand a couple of "Oops, I left my lights on. Can you give me a boost?" mistakes before it is destroyed, whereas a deep-cycle battery may be subjected to much higher levels of cycling.

Depth of Discharge %	Specific Gravity @ 75° F (25° C)	Cell Voltage
0	1.265	2.100
10	1.250	2.090
20	1.235	2.075
30	1.220	2.060
40	1.205	2.045
50	1.190	2.030
60	1.175	2.015
70	1.160	2.000
80	1.145	1.985
90	1.140	1.825
100	1.130	1.750

Table 10-1. Using the hydrometer shown In Figure 10-4, it is possible to accurately determine the state of charge, specific gravity, and voltage of each cell in a lead-acid battery bank.

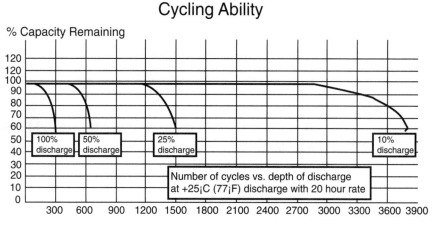

Cycling Ability

% Capacity Remaining

CAPACITY WITHDRAWN	CYCLES
TYPICAL CYCLING PERFORMANCE*	
100%	300
50%	650
25%	1,500
10%	3,800

* Dependant upon proper charging and ambient temperatures

Figure 10-5. This graph relates the life expectancy in charge/discharge/charge cycles to the depth of discharge level. Grid-interconnected systems do not generally cycle deeply or frequently unless the electrical utility is particularly unreliable in your area.

Figure 10-5 graphs the relationship between the life of a battery in charge/discharge/charge cycles and the amount of energy that is taken from the cell. For example, a battery that is repeatedly discharged completely (100% depth of discharge) will only last 300 cycles. On the other hand, if the same battery is cycled to only 25% depth of discharge, the battery will last 1,500 cycles. It stands to reason that a bigger battery bank will provide longer life, albeit at a higher cost for the added capacity. A good level of depth of discharge to shoot for is a maximum of 50%, with typical levels of between 20% and 30%.

A grid-interconnected battery system spends the majority of its life filled to capacity waiting for the next utility failure. Provided power outages are not an everyday occurrence, frequent cycling should not be a major concern. On the other hand, depth of discharge considerations may become a major issue where grid interruptions are frequent. When a lengthy power outage occurs, significant battery-capacity depletion can result from inexperienced operation. Novice operators should consider battery-metering equipment as discussed later in this chapter.

Operating Temperature

A battery is typically rated at a standard temperature of 75°F (25°C). As the temperature drops, the capacity of the battery drops as a result of the lower "activity" of the molecules making up the electrolyte. The graph in Figure 10-6 shows the relationship between temperature and battery capacity. For example, a battery rated at 1,000 amp-hours (Ah) at room temperature would have its capacity reduced to 70% at -4°F (-20°C), resulting in a maximum capacity of 700 Ah. If this battery is to be stored outside where winter temperatures may reach this level, the reduction in capacity must be taken into account. Installing batteries directly on a cold, uninsulated cement floor will also cause a reduction in capacity for the same reason. Be sure to use a wooden skid or other frame to allow room-temperature air to circulate around the battery, maintaining an even temperature.

Freezing is another concern at low operating temperatures. A fully charged battery has an electrolyte-specific gravity of approximately 1.25 or higher. At this level of electrolyte acid strength, a battery will not freeze. As the battery becomes progressively discharged, the specific gravity gradually falls until the electrolyte resembles water at a reading of 1.00.

For cold-weather applications you may consider a Nickel Metal Cadmium or NiCad battery. Although more expensive than standard lead-acid batteries, the NiCad cell is relatively unaffected by extremes in temperature,

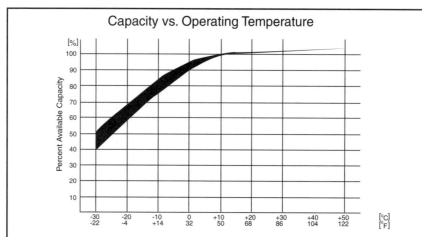

Figure 10-6. This graph shows the resulting reduction in battery capacity as the ambient temperature is lowered. A thermometer should be used to correct the specific gravity reading of very cold or overly hot electrolyte.

particularly where deep-discharge cycles and cold weather threaten to destroy lead-acid varieties.

Battery Sizing

A single cell does not have enough voltage or capacity to perform useful work. As discussed in Chapter 1.3, single cells may be wired in series to increase voltage and/or in parallel to increase capacity.

Batteries are just big storage buckets. Pour in some electrons and the "buckets" will fill with electricity. If cells are wired in parallel, the capacity is increased. If the cells are wired in series, the voltage or pressure rises.

Batteries may be purchased in several voltages. Individual cells have a nominal rating of 2 V for lead-acid and 1.2 V for NiCad brands. The lead-acid battery shown in the background of Figure 10-2 has two cells, each of which can be identified by the servicing cap

Figure 10-7. For applications where the battery bank must be stored outdoors in extremely cold locations, consider the more expensive yet longer life NiCad style. (Courtesy Saft Battery Company)

on the top. Using interconnecting plates inside the battery case, the cells are wired in series, providing 2 + 2 volts = 4 V total capacity. The batteries in the foreground have three cells, providing an output of 6 V. Using the same approach, a car battery contains six cells, providing a nominal 12 V rating. Typical battery voltages for grid-interconnected renewable energy systems are 24 V and 48 V.

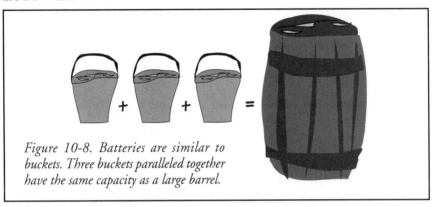

Figure 10-8. Batteries are similar to buckets. Three buckets paralleled together have the same capacity as a large barrel.

It stands to reason that the more energy consumed, the larger the storage facility required to hold all the electricity. While the capacity of buckets is given in gallons or liters, the capacity of batteries is rated using watt-hours or amp-hours, units of electrical energy. We discussed earlier that a battery's voltage tends to be a bit elastic, with the voltage rising and falling as a function of the battery's state of charge (see Table 10-1). Additionally, when the battery is charged from a source of higher voltage such as a PV array in full sun, the voltage increases further. It is not uncommon for a battery bank with a nominal 24 V rating to have a voltage reading of 30 V when undergoing equalization charging.

This fluctuation of battery voltage makes it very difficult to calculate energy ratings. You will recall that energy is the voltage (pressure) multiplied by the current (flow) of electrons in a circuit multiplied by the amount of time the current is flowing. The problem with this calculation when it is applied to a battery is deciding which voltage to use.

Energy (watt-hours) = Voltage x Current x Time

Battery manufacturers are a smart bunch. To eliminate any confusion with ratings, they have simply dropped the voltage from the energy calculation, leaving us with a rating calculated by multiplying current x time.

Battery Capacity (amp-hours) = Current x Time

The assumption is that there is no point in trying to shoot a moving targer. If we really want to look at our energy storage in more familiar watt-hour terms, we can simply multiply the amp-hour rating by the nominal battery-bank voltage.

Let's look at how this relates to the storage level our battery must have in order to run the essential loads of our home when the grid goes south. If you haven't already done so, it will be necessary to complete the *Energy Sizing Worksheet* in Appendix 7. This form will determine the amount of electrical energy you require in watt-hours per day to operate these essential loads.

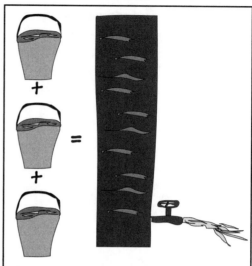

Figure 10-9. Batteries wired in series increase the voltage or pressure. In the same manner, a tall tank will shoot water further than a short tank due to the higher pressure.

For our example, we will assume an energy consumption of 4,000 watt-hours (4 kWh) per day for a 24-hour blackout. This is a reasonable amount of energy for grid-interconnected system users who wish to ride through power failures. In this case, it will be necessary to provide a separate wiring circuit for essential loads such as house lights, a refrigerator, and a few plugs to receive power during a grid interruption. Large or inefficient loads such as electric heating or air conditioning would remain off during the power failure. A detailed discussion of this system configuration follows in Chapter 14.

Remember that we have to take into account the usable amount of energy stored in the battery bank, not the gross or rated amount. A maximum depth of discharge is typically 50%, with less being recommended.

4 kWh capacity required x battery derating factor of 2 = 8 kWh capacity

Some manufacturers will rate the capacity of their batteries in watt-hours or kilowatt-hours; however, as we discussed earlier, most choose to use amp-hours. To convert watt-hours (kWh) to amp-hours (Ah), divide the desired battery rating by the nominal battery voltage required for your system, typically 24 or 48 Volts (assume 24 V).

8 kWh capacity (8,000 Wh) ÷ 24 V = 333.33 Ah capacity
333 Ah capacity

A quick scan of the resource guide data in Appendix 3 indicates that there are no batteries available with exactly 333 amp-hours of capacity. Remember, however, that batteries can be connected in series to increase voltage and in parallel to increase capacity. Two sets of 150 Ah batteries wired in parallel provide 300 Ah of capacity. This is a bit below the calculated rating, but it's close enough. We could also consider 2 sets of 200 Ah batteries, offering 400 Ah. A bit of juggling may be required, but either way we're in the ballpark.

Figure 10-10. This compact battery bank takes up a small amount of room in the utility closet. With a rating of 24 volts at 1,300 amp-hours of capacity, it gives the owners of this home enough energy to ride through several days of grid interruption.

As discussed earlier, under-sizing the battery can lead to over-discharging and shortening the battery's life. If you have to estimate and select sizes, it is always in your best interests to round up. Battery capacity is like money in your bank account: it never hurts to have extra on hand.

Each of the two rows of eight batteries shown in Figure 10-10 is shown in the schematic in Figure 10-11. The columns comprise 3 batteries. Each battery has two cells, as you can see from the two servicing caps on their cases. The nominal voltage of each battery is 4 V (2 cells x 2 Volts/cell). The batteries are wired in series by connecting successive "-" to "+" terminals, making the voltages addi-

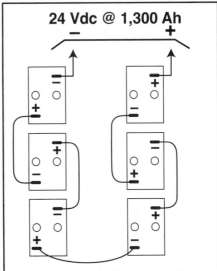

Figure 10-11. The batteries in Figure 10-10 are shown here in a schematic. Each of the six batteries is wired in series, resulting in increased voltage.

tive. Thus, six batteries wired in series multiplied by 4 V per battery provide a total bank of 24 V.

Each of the batteries shown in Figure 10-10 has a capacity of 1,300 Ah. Wiring the batteries in series does not increase the capacity. In order to increase capacity, the batteries must be connected in parallel so that both banks are forcing electrons into the circuit at the same time. (To help visualize this concept, picture a car stuck in a snowbank. One person pushing the car may not provide enough energy to get the vehicle out, while a second person giving a hand may easily free it.) The parallel connection of multiple strings of batteries will be discussed further in Chapter 14, "Putting it All Together"

Hydrogen Gas Production

Connecting a battery to a source of voltage will push electrons into the cells, recharging them to a full state of charge. As a liquid-electrolyte battery approaches the fully charged state it will be unable to absorb additional electrons, causing the electrolyte to start bubbling and emit hydrogen gas. "Maintenance-free" gelled-cell batteries are designed to minimize the possibility of hydrogen buildup.

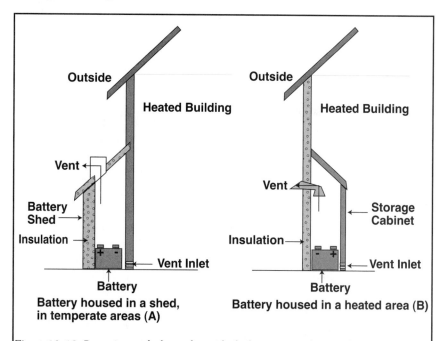

Figure 10-12. Batteries can be located outside the living area where ambient temperatures do not deplete storage capacity (A). For colder climates the batteries may be stored in a sealed box or closet (B). Both installations require a vent with bug-screen directed outdoors.

Everyone knows the story of the Hindenburg dirigible: a few sparks and the hydrogen gas exploded into a ball of fire. Preventing this in a renewable energy system is actually fairly easy.

The battery bank should be accessible to allow for servicing and periodic inspection. However, safety is of primary importance. Storing batteries in a locked closet, shed, or cabinet is a wise decision. The storage closet may be vented with a simple pipe that can be made with a dryer vent or plumbing pipe (2"/50 mm) in diameter with screening over the exterior opening. Make sure the pipe slopes slightly upwards from the battery room to the outside as hydrogen is lighter than air and will rise on its way outdoors.

For rooms with long or serpentine vent shafts, a power-venting fan can be installed as shown in Figure 10-13. Pressurizing fans are operated by a voltage-controlled switch contained within the inverter; the voltage indicates the state of charge and hence hydrogen production, activating the fans.

An alternate method of controlling hydrogen gassing is to install a set of reformer caps (Hydrocap Corporation) such as those shown in Figure 10-14. These caps contain a catalyst that converts the hydrogen and oxygen by-products of battery charging into water

Figure 10-13. This battery room is power vented. Installing small computer fans in the wall allows the room to be pressurized, forcing hydrogen outside.

Figure 10-14. HydroCaps replace the standard vent caps on each battery cell. A special catalyzing material converts hydrogen and oxygen byproducts back into water.

Figure 10-15. In temperate climates batteries may be stored in suitable outdoor cabinets, eliminating any concerns regarding hydrogen production. (PSR cabinet courtesy Outback Power Systems Inc.)

that simply drips back into the cell. The net effect is eliminated hydrogen gassing and lower battery water usage. Note, however, that hydrogen reformer caps control but do not completely eliminate hydrogen gassing, making some supplementary ventilation necessary.

Safe Installation of Batteries

Large deep-cycle batteries are heavy and awkward. Some models weigh up to 300 lbs (136 kg) each. It is important that any frames, shelves, and mounting hardware be of sufficient strength to hold them securely. During installation, make sure you have adequate manpower to lift and place the batteries without undue straining. Tipping a battery and spilling liquid electrolyte (acid) all over you is very dangerous. Follow these rules to ensure a proper and safe installation (See Figure 10-19):

- Batteries must be installed in a well-insulated and sealed room, closet, or cabinet, preferably locked. There is an enormous amount of energy stored in the batteries. Children (or curious adults) should not be allowed near them.
- Remove all jewelry, watches, and metal or conductive articles. A spanner wrench or socket set accidentally placed across two battery terminals will immediately weld in place and turn red-hot. This can cause battery damage, possible explosion, and severe burns.
- Hand tools should be wrapped in electrical tape or be sufficiently insulated so as not to come in contact with live battery terminals.
- All benches, shelves, or support structures must be strong enough to carry the massive weight of the battery bank. The batteries shown in

Figure 10-10 weigh 900 lbs (400 kg), so use extreme caution when handling.

- Wear eye-splash protection and rubber gloves when working with batteries; splashed electrolyte can cause blindness. Also, wear old clothes or coveralls, since electrolyte just loves to eat your $80 jeans.
- Keep a 5 lb (2.3 kg) can of baking soda on hand for any small electrolyte spills. Immediately dusting the spilled electrolyte with baking soda will cause aggressive fizzing, neutralizing the acid and turning it into water. Continue adding soda until the fizzing stops.

Battery and Energy Metering

Electricity is invisible, making it difficult to determine how much energy is actually stored in your battery bank. While it is possible to use nothing more than a hydrometer to measure the stored energy level, this requires a fair bit of tedious fiddling and working with corrosive acid. A better approach is to install a multifunction meter such as the model shown in Figure 10-17.

These meters operate much like an automobile gas

Figure 10-16. This compact battery bank takes up a small amount of room in a utility closet or other out-of-the-way area in the home. (Courtesy Outback Power Systems Inc.)

gauge, indicating how full or empty the tank is. Most meters provide dozens of features, the most important being battery voltage measurement, energy capacity in amp-hours and a measurement of the current that is flowing into (charge) or out of (discharge) the battery. These meters can be mounted a long distance from the battery bank, enabling you to read capacity from a convenient location in the house rather than running to the garage or basement.

Meters perform their work through a connection with a device known as a "shunt" (see Figure 10-18), a high-capacity resistor that either bypasses a small amount of the current flowing into or out of the battery or shunts it to the meter.. As the circuit is DC-rated, the direction of current flow is easily recorded. The amount of energy shunted is a calibrated ratio of the total amount of energy flowing in the battery circuit. Increasing the current flow into or out of the battery circuit causes a calibrated current to flow to the meter.

Figure 10-17. A multifunction meter such as this Trace model by Xantrex (www.xantrex.com) takes most of the guesswork out of battery status monitoring. (Courtesy Xantrex Technology Inc.)

By constantly monitoring the flow into and out of the battery, the meter keeps track of the battery energy level in amp-hours and percentage full. Or it almost keeps track. While there is no doubt that energy meters are very useful and accurate devices, they are not perfect. Batteries are not perfect either. The process of converting electricity into chemical storage within the electrolyte is not 100% efficient. During the charge and discharge cycling of the battery, some of the electrical energy we put in is lost due to conversion inefficiencies. Worse yet, this inefficiency is not the same for adjacent cells, nor does the inefficiency level remain constant over the life of the cell. As a result, the energy meter will require recalibration periodically when the battery is at a known fully charged state (after measurement with a hydrometer).

Electrolyte specific gravity and battery voltage are directly related:

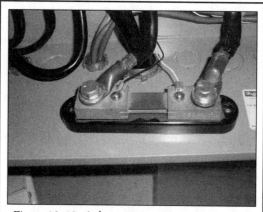

Figure 10-18. A shunt is a precision resistor wired in series with the negative lead of the battery. A small current in proportion to the battery charge or load current is "shunted" to a multifunction meter such as the one shown in Figure 10-17.

Battery Voltage = Electrolyte specific gravity + 0.84

For this equation and the hydrometer reading to be accurate, however, the batteries should be under light or no load, with the electrolyte near room temperature. It is also important to perform this measurement approximately two hours after charging the batteries to ensure that gas bubbles in the electrolyte do not lower the specific gravity reading.

The only way of absolutely knowing the energy level and general "health" of a battery is to use a hydrometer and measure the specific gravity. A thermometer should be used to correct the specific gravity reading of very cold or overly hot electrolyte according to the manufacturer's ratings. Correlate the corrected specific gravity to the data in Table 10-1 to determine the actual state of charge.

Batteries are charged to a known level a few times a year by performing an equalization charge using grid or renewable energy. During this process the batteries will reach their known fully charged state and the meter may be reset, calibrating the battery level with the meter reading for the next couple of months. Equalization charging will be covered in Chapter 15 – "Living with Renewable Energy".

Figure 10-19. Battery installations should be neat and tidy and performed with the necessary tools on hand to measure electrolyte, specific gravity, and temperature. A clipboard and written record of each cell's "health" will ensure long battery life.

In addition to the "main" features described above, metering may also contain some or all of the following "secondary" indicators:

- number of days since batteries were fully charged
- total number of amp-hours charged/discharged since installation
- time to recharge battery
- voltage too low for proper operation
- battery charging
- battery fully charged

Summary

In this chapter we have discussed battery selection and safe installation. Chapter 11 – "DC Voltage Regulation" discusses the equipment necessary to ensure proper charging and regulation of the battery bank. Future chapters deal with the specifics of interconnecting cells, safe wiring practices, and maintenance. After all, when the grid goes down, your home should be a shining beacon of bragging rights for all to see.

11
DC Voltage Regulation

A charge controller is an important element of a renewable energy electrical system equipped with batteries and is also essential for some models of wind and hydro turbines. Charge controllers are designed to provide protective functions, ensuring that batteries are not under- or overcharged, thereby prolonging battery life. Wind and hydro turbines may require an electrical load at all times. If your system is configured in this way, a diversion load controller will provide a constant load in the event of grid or inverter failure.

Grid-interconnected systems not equipped with batteries or these types of turbines do not require any form of charge controller, as all of the available electrical energy is pumped directly into the grid. If your system is grid-connected *without batteries*, or if you are using PV only, feel free to skip this chapter.

All renewable energy systems sometimes generate more energy than can be reasonably used in the home. Even small configurations will generate excess energy if no one is there to use this energy. Electricity generated by the sun or wind is converted into alternating current using a device known as an inverter. The alternating current is then supplied to your home appliances, with any excess being fed into the electrical grid for sale to the utility.

Should the electrical main supply fail, in the case of a *grid-dependent* design the renewable generating source and inverter will disconnect and no further energy will be produced for the house or exported for sale.

Systems equipped with a backup battery bank are known as *grid-interactive* systems and are designed to ride through short-term blackouts. In this case, charge controllers form an integral part of the system design.

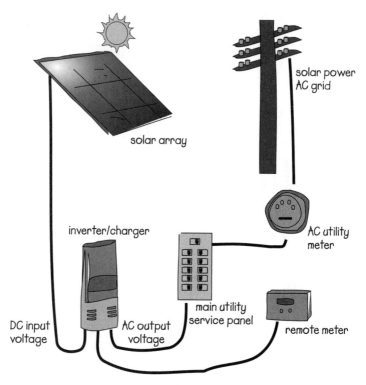

Figure 11-1. The majority of urban, grid-interconnected renewable energy systems are not equipped with battery backup. If your system is configured this way, you do not require any DC voltage regulation and can skip to the next chapter.

As a battery's state of charge increases, its voltage also increases. The renewable energy source produces power at a high voltage in order to provide sufficient "pressure" to cause electrons to flow into the battery. We learned in Chapter 8, "Photovoltaic Electricity Generation," that a typical PV cell will produce 17 volts (V) for a 12 V nominal battery system. The 5 V difference in pressure allows electrons to flow from the higher voltage source to the lower one, providing the charging current to the battery bank.

As the battery bank "fills," its voltage rises. Upon nearing the fully charged state, the battery will start outgassing hydrogen and oxygen gas as well as heat. This condition is known as overcharging, and if it continues for prolonged periods damage to the battery will occur. This is one example of where a charge controller is required.

Charge controllers are available in two distinct categories known as *series controllers* and *diversion controllers*. Small off-grid systems and the majority of urban grid-interconnected designs use the series control method. Large, off-grid systems, particularly those utilizing wind and micro-hydro sources, use the diversion control method.

Series Controller

As the name implies, the series controller is wired in series between the PV array and the battery bank as shown in Figure 11-3. In this arrangement, the voltage output from the PV array is fed through the series-connected charge controller prior to being supplied to

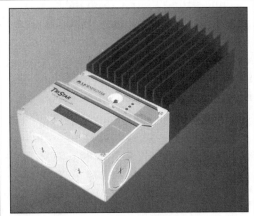

Figure 11-2. The charge controller ensures that your battery bank is properly charged and operating within given operating parameters. (Courtesy Morningstar Corporation)

the battery. The charge controller monitors the battery voltage, and provided it is below a fully charged condition, PV power is allowed to flow into the battery. When the battery voltage level is sufficiently high to fully charge the battery, the charge controller will taper charging current or completely

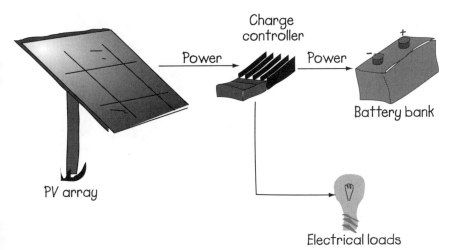

Figure 11-3. The series-connected charge controller is able to connect and disconnect the PV array from the battery. This action limits charging current and voltage to safe levels, prolonging battery life.

disconnect the PV array from the battery, slowing or stopping the charging cycle. As the battery is subjected to household electrical loading, the charge controller senses the drop in battery voltage as energy is removed from the cells. The controller reconnects the PV array, restarting the charging process. Think of the series controller as an automatic light switch that turns the flow of current to the batteries on and off depending on whether the batteries require additional energy.

During the night, the charge controller performs no function other than to electrically isolate the battery bank from the sleeping PV array. This is prevents energy from flowing *backwards* from the batteries into the array, which can occur at night when the battery voltage is at a higher level than the voltage of the PV array. The large surface area of the PV array may absorb a small amount of energy from the battery and dissipate it as heat. Over the course of a long winter night, this energy loss is measurable and can contribute enough inefficiency to become a concern.

For this reason, most series charge controllers contain a "PV array nighttime reverse-current protection" feature to eliminate the possibility of electricity backflow.

Maximum Power Point Tracking

A word is required about *maximum power point trackers* or MPPT controllers. This technology, which is relatively new to the renewable energy market, is now available in charge controllers manufactured by several companies. All MPPT charge controllers provide the series charge control functions described above. Where these devices surpass standard controllers is in their ability to significantly improve the overall efficiency and power output of a renewable energy source, in essence producing more energy dollars from the same amount of equipment.

Figure 11-4. This charge controller is rated for 60 Amps of continuous load and is complemented with Maximum Power Point Tracking capability. (Courtesy Outback Power Systems)

The MPPT function is a little bit difficult to understand but well worth a few moments to review. The power of a PV module is the product of the voltage and the current. This can be expressed by multiplying the rated current by the rated voltage under load, resulting in the rated power of the module, shown in graphical form in Figure 11-5. The output voltage of a given module is shown on the "x" axis with the output current on the "y" axis. In Chapter 8 we discussed the fact that a PV module generates a very high open circuit voltage when not connected to a load (i.e. when no current is flowing). This point on the graph is position Voc (Voltage Open Circuit), where zero current is flowing and no work is being done.

As a load is connected to the panel, increasing current causes the voltage to slowly drop. A short circuit (directly connecting the "+" and "-" terminals) will result in high current levels, but the voltage will approach zero, thereby producing no useful power and again, no work can be done.

The most efficient point on the curve occurs when the product of the voltage pushing and the current flowing produce the highest value. This occurs at one point, which is known as the Maximum Power Point (MPP). Unfortunately, our battery or utility grid voltage is unlikely to be located at this point. As we have learned, battery voltage is quite elastic, so there is a high probability that the PV module output will only transit this point occasionally.

The MPPT controller monitors the PV panel

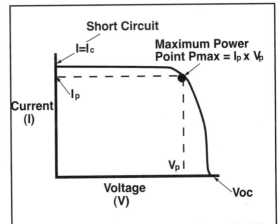

Figure 11-5. The Maximum Power Point (MPP) for a PV module is the point at which current and voltage outputs provide the maximum amount of power (Power = Voltage x Current).

or wind turbine output power in relation to battery or inverter load and determines where the MPP is for the level of light illuminating an array or wind powering a turbine.

Testing shows that MPPT systems increase the wattage of a PV array by an average of 15% to 30%. For a PV array pushing 800 W into the utility grid or battery bank, this will provide an average increase of 120 W with a peak of double this figure. This is approximately equal to one or two modules of "free" power and more dollars from the sale of electrical energy.

For PV-only systems, consider an MPPT controller as an alternative or supplement to an active tracking mount to increase daily energy production.

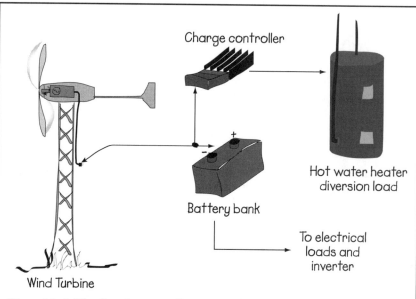

Figure 11-6. The diversion controller strategy shunts or diverts excess energy to an auxiliary air or water heater unit according to battery voltage or state of charge.

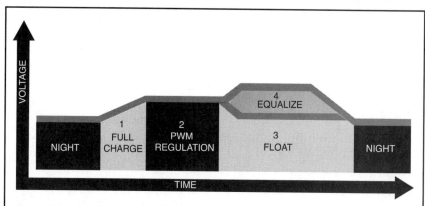

Figure 11-7. A charge controller regulates the amount of energy flowing into the batteries, ensuring a proper charge cycle. Quality charge controllers offer several charging modes. (Courtesy Morningstar Corporation)

Diversion Charge Controller

The diversion controller method (see Figure 11-6) is used primarily for larger off-grid systems or grid-interconnected designs that incorporate wind or micro-hydro turbine generators. In this configuration, the renewable energy source is connected either to a battery bank or to an inverter. All of the energy produced by the turbine flows into the battery or inverter, providing a load connection at all times. This is an important distinction between the two designs of charge controller. Where a PV array can be connected and disconnected at will, many wind and hydro turbines must have an electrical load connected at all times. Removal of the load during operation can cause the turbine to accelerate to a speed where damage can occur. This is similar to a car engine: driving while holding "the pedal to the metal" results in speeding tickets but no engine damage; however, doing the same with the car in neutral will almost certainly destroy the engine, as there is no load to limit its speed.

To ensure a constant load for the turbine without overcharging the battery, the controller will "shunt" or divert excess energy to the diversion load. The diversion load in Figure 11-6 is a water heater, although an air heating or other dump load can be used. With this system, the diversion controller is able to divert all or part of the turbine's energy depending on the battery bank voltage and state of charge, while maintaining a proper load level for the renewable source.

Wind and hydro turbines that are grid dependent do not have a battery bank. They require a constant load and use the diversion controller in a slightly different manner. In this configuration, the turbine supplies power to both the inverter and the diversion controller simultaneously.

When the grid is within its normal operating parameters, the diversion controller is programmed to allow all of the generated energy to flow into the inverter, in turn feeding the house and grid. When the grid fails and the inverter enters standby mode the output voltage of the turbine will rise as a result of the lack of load. The controller senses this increase in voltage and shunts the excess energy to the diversion load.

Voltage Regulator Selection

Regardless of which voltage regulator design is selected, it is important to match its electrical rating to your system. The regulator rating is based on the maximum current that is allowed to flow into the batteries (series regulator) or into the diversion load (diversion regulator). It will be necessary to add the peak charging/diversion currents expected in your system. Once this value is determined a derating or safety factor of 25% is normally added. If your charge or diversion loads require a higher capacity rating, multiple controllers may be wired in parallel to divide the electrical load.

peak charging or diversion current x 25% safety factor = regulator rating in Amps

For further details refer to Chapter 14, "Putting it All Together Safely".

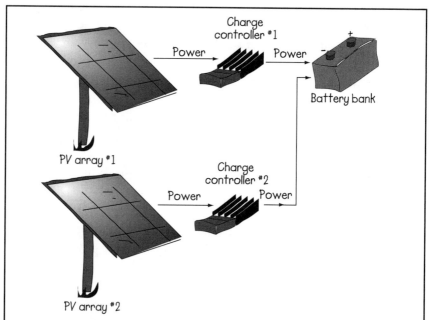

Figure 11-8. If the power rating of your PV panel exceeds the rating of the charge controller, connect multiple units in parallel as shown. The PV array does not have to be physically split into two groups, but can simply be wired to divide the power output so that it is within the maximum safe operating range as recommended by the controller manufacturer.

Charging Batteries from the Grid

Grid-interconnected systems are often supplied with inverters capable of drawing electricity from the grid to charge the battery bank. The inverter's built-in charge control circuits will work in parallel with the renewable source, ensuring that the battery bank is always fully charged and ready the next time a blackout hits.

Charging Strategy (all controller configurations)

When the voltage is below a level that indicates the battery is not fully charged, all of the power from the renewable source is allowed to pour in. As the battery voltage rises, the series charge controller rapidly cycles on and off (several hundred times per second), pulsing the generated power into the battery. In a similar manner the diversion controller rapidly cycles power to the diversion load. The net effect is to direct energy in a controlled manner, maintaining the battery voltage within preset limits.

To visualize this concept, go into a darkened room and flip on an incandescent lamp. The room will be lit at maximum brightness. If you rapidly flip the switch on and off, the bulb will alternate between states of full and zero brightness. The effect on your eyes (notwithstanding the flicker) is that the room is seen in a dimmer light. If the light switch could be controlled so that the switch spent more time on than off, the brightness would increase. Likewise, if the switch spent more time off than on, the brightness would decrease. Engineers term this rapid on-and-off switching *Pulse Width Modulation* or PWM.

To understand how the PWM concept is incorporated into a charge control system, let's take a look at a typical twenty-four-hour day in the life of a battery bank during a grid blackout, assuming that PV arrays are used for daily charging. Figure 11-7 shows the effect of battery voltage over one complete night/day/night cycle. The graph shows time on the horizontal or "x" axis, while battery voltage is shown on the vertical or "y" axis.

Nighttime
Starting at the left-hand side of the graph, the battery voltage is shown at its lowest state at night. During the night lights are used, the fridge operates, and other loads draw energy from the battery bank. As we learned in Chapter 10, "Battery Selection and Design", the battery voltage will decrease as energy is removed from the battery. During this period, the charge controller prevents electrical backflow (discussed earlier) and waits patiently for the sun to rise.

Sunrise – Full or Bulk Charge
As the sun rises and shines on the PV array, electrical energy will start flowing and begin charging the battery. The battery voltage will start to rise according to the battery state of charge, as discussed in Chapter 10.

Until the battery state of charge reaches approximately 80%, all of the PV array power is applied directly to the battery, as shown in stage (1) of Figure 11-7. This charging condition is called "full or bulk charge" mode. It terminates when the battery voltage rises to 14.6 V for a nominal 12 V battery bank (consult manufacturers' data sheets for exact charger settings). For 24 V or 48 V battery banks, multiply the described settings by two or four respectively.

Absorption or Tapering Charge

When the battery reaches an approximately 80%-full state of charge, some of the energy is wasted through "boiling" and out-gassing. Boiling is a slang term used to describe the breakdown of water in the battery electrolyte into its component elements, hydrogen and oxygen. This occurs when there is more energy applied to the battery than it can absorb.

Hydrogen is a gas that is lighter than air and very explosive when mixed with oxygen and an accidental spark, open flame, or cigarette. For this reason, batteries should be stored in locked and ventilated cabinets. It is possible to reduce hydrogen production while providing the optimum charging current to the battery by introducing a PWM tapering current, as shown in stage (2) of Figure 11-7.

During this stage, charging current is automatically lowered in an attempt to maintain the battery voltage at the bulk setpoint level of 14.6 V. Absorption charging continues for a defined period of time, which is typically one to two hours.

The excess energy produced during the absorption charging stage, which can be a considerable amount, is either wasted or diverted depending on the type of charge controller installed. A series controller will waste the excess energy. As this energy is non-polluting and often small in quantity, there is no concern. A diversion controller will shunt or redirect the excess energy to a dump load such as an electrical water or air heater, increasing the system's efficiency. (Remember that all of this wasting and diverting occurs only during times when the grid is "down" and the house is operating on battery backup).

Float Stage

Once the battery is fully charged, the charge controller will reduce the applied voltage to 13.4 V and begin float mode, as shown in stage (3) of Figure 11-7. When the battery reaches this stage, very little energy is flowing into it and nearly 100% of the renewable energy will be wasted or shunted to the diversion load. Float mode will remain in effect until the battery voltage dips below a preset amount (typically 80% of full state of charge) or until the renewable energy source stops producing power at night. The process is then repeated for the next cycle.

Equalization Mode

Caution!

Sealed or "maintenance free" batteries do not require equalization charging. Placing them in such a condition can result in fire or cause the battery to explode.

To visualize the equalization process, think of batteries containing electrons as buckets containing water, with one bucket equivalent to one battery cell.

Over the course of a few dozen charge cycles, energy is taken out of and replaced into the battery. This is the same as if I asked you to take four cups (one liter) of water out of each "cell" or bucket, representing a day's electrical load consumption. Assume that our PV array produces enough energy to replace three cups (750 ml) of "energy" into the cells, leaving us with a deficiency of one cup (250 ml). This process is repeated with different amounts of water being added or removed over the course of a month or two, until such time as the batteries are back to a full state of charge. In our example, the buckets are also full.

In reality, the water in each of the buckets would not be at exactly the same level. A certain amount of spillage and uneven amounts of water removal will leave the buckets with varying amounts of water in them. This is the exact scenario that plays out in your battery bank. Over time there is a gradual change in the state of charge in the cells, which can be determined by comparing the specific gravity of one cell with that of adjoining cells.

Figure 11-9. Using large amounts of diverted electrical energy is easy if heat enters the equation. An air heating element such as the model shown in this photo (available from www.realgoods.com) will absorb up to 1 kW of excess energy and provide some home heating to boot.

If the difference is allowed to continue for an extended period, the cell with the lower state of charge can fail prematurely and the amount of available energy stored in the battery bank can be reduced.

To correct this situation, a periodic (typically once per month) controlled overcharge is conducted. Known as an equalization charge, this process is illustrated in stage (4) of Figure 11-7. Chapter 15, "Living with Renewable Energy", discusses how to determine when equalization is required.

Equalization charging is normally conducted early in the morning on a sunny or windy day. The charge controller is set to equalization mode and the normal bulk cycle (step 1) is completed. Upon entering equalization mode the controller raises the battery voltage to 15.5 V (step 4). Equalization mode is maintained for approximately two hours, after which battery voltage is reduced and the charge controller automatically enters float mode (step 3).

If we go back to our bucket example, the effect of equalization is similar to using a garden hose to add water to the buckets and deliberately overfilling them. When you stop adding water (equalization completed) the buckets are topped up right to the rim.

Where does the "extra" electricity go during the equalization mode?

Figure 11-10. This home uses a high-efficiency, in-line gas water heater. The conventional storage water heater is the diversion load. It absorbs waste energy by preheating the cold water fed to the gas water heater.

During this charging stage, the excess energy applied to the batteries will cause violent bubbling of the electrolyte, producing large amounts of hydrogen and oxygen gas as well as heat and water vapor. It may be necessary to monitor electrolyte levels and battery temperature during this charging stage to ensure that battery parameters are within manufacturer ratings.

If your batteries are equipped with hydrogen reformer caps, ensure that they are removed during equalization mode.

Diversion Loads

Wind and hydro turbines normally require that an electrical load be connected at all times. Energy that is generated by these sources and delivered to the grid during periods of blackout must go somewhere. In addition, a grid-interactive system (one equipped with battery backup) requires a constant electrical load because the charge controller diverts excess energy away from the fully charged batteries. This is where the diversion load comes into play.

A diversion load is any load that is large enough to accept the full power of a renewable source. Although the diversion load may simply waste the excess energy applied to it, a better approach is to put this juice to work. After all, you did pay for the wind turbine, so why waste the "excess" energy other people have to pay for? Suppose you have a 2 kW wind turbine which operates at maximum output for 5 hours during a grid failure while you and your family are on vacation.

2 kW wind turbine x 5 hours of operation = 10,000 Watt-hour production

Assume for a moment that the house loads are zero (no fridge or lights on). What can you do with this energy? 10 kWh of production is a lot of energy. You could operate a thousand 10 W compact fluorescent lamps for one hour or ask your neighbors over for a really big party at your house.

This free energy can be used in many ways to help offset other systems that cost money to operate. A common solution is to use the excess energy for space and water heating. If all of your waste energy is produced during the swimming season, consider dumping the energy into a hot tub or swimming pool. Just remember that this load must not be considered part of your overall energy budget because the diversion load will only be active when the grid has failed.

The most common diversion loads are air and water heaters such as those shown in Figures 11-9 and 11-10. For PV-based systems, the majority of excess energy is produced during the summer months when air heating is not required. Wind systems tend to provide maximum power during the late

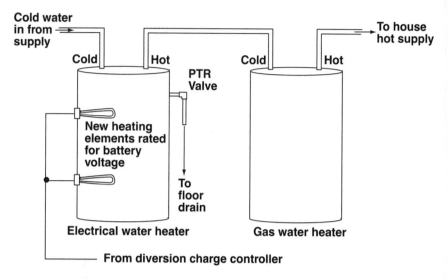

Figure 11-11. A diversion water heating load is connected in series with a standard gas water heater. The elements of the water heater are replaced with heating elements that have the same nominal voltage as the battery bank. They are connected to the diversion charge controller.

winter and spring periods. Consider the choice carefully. No one wants an air heater cooking when the mercury is in the 90s. My personal preference is a water-heating load such as the one in Figure 11-10. This arrangement requires a bit of installation work, but the excess energy will offset water heating fuel costs when the grid goes out.

In this system, a standard electric storage water heater is purchased and the 240 V heating elements are removed. New elements are installed which have the same voltage rating as your battery bank or your wind or hydro turbine. The elements are then wired to the diversion-type charge controller. The water heater is plumbed so that cold water flows into the electric heater and out to the standard model, as shown in Figure 11-11.

During normal operation, the cold water supply enters the electric water heater. If there is no excess energy, the cold water will absorb some room heat, capturing a small amount of supplementary energy before heading to the regular water heater. As this incoming cold water is below the setpoint temperature, the regular heater will supply the energy necessary to meet demand.

During periods of excess energy production resulting from grid failure, the diversion charge controller supplies the low-voltage electric water heater.

This energy may heat the water to the desired setpoint temperature or beyond. Feeding preheated water into the regular water heater reduces or eliminates the need for any further heating, reducing your purchased energy requirements and saving you dollars.

It is possible for the temperature of the water in the diversion electric heater to rise above the setpoint temperature, particularly during travel or vacation periods which often occur in the energy-rich summer months. Safety (as well as building codes) dictates that a Pressure and Temperature Relief (PTR) valve be installed on the tank. The outlet pipe should be run to a floor drain or other suitable exhaust to allow very hot water to be safely drained away in the event that the water tank temperature rises to an unsafe level.

For the same reason, your plumber should install a buffering valve. These valves ensure that the hot water supply delivered to the plumbing fixtures is within a comfortable and safe temperature range.

For people considering the addition of a solar water heating system, diversion heating elements may be combined with the required water storage tank.

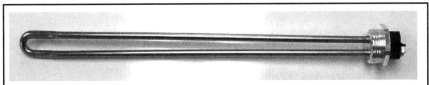

Figure 11-12. Alternate voltage water heating elements are available from numerous sources, including Real Goods at <u>www.realgoods.com</u>.

12
DC to AC Conversion Using Inverters

I n earlier chapters, we dealt with renewable energy sources that produce electrical energy in direct current (DC) from a PV array. Wind and hydro turbines generate alternating current (AC), but convert it back to DC for storage in battery banks.

This chapter explains how to convert DC power back to AC power that can be used for electrical appliances and can also be sold to the electrical utility in grid-interconnected systems.

In the "old days," if you wanted to build a renewable energy system you could only do it using 12 Volt (V) direct current appliances similar to those used in early recreational vehicles. Grid-interconnection of renewable energy was as

Figure 12-1. The renewable energy world would be lost without utility-grade sine wave inverters and controls such as this. (Courtesy Outback Power Systems)

far away as the idea of the Internet. Times change, the Internet is here, and so are very high quality inverters for both on- and off-grid applications.

The basic inverter is a device that takes low-voltage DC power (from a battery bank or directly from another energy source such as a PV panel or wind turbine) and converts it to alternating current. It then "steps up" the voltage to match domestically supplied power from your utility. In practice, many inverters offer a whole host of additional features and functions:

- battery charging capability
- transferability of house power between a generator and the inverter
- low- and high-voltage alarms and disconnection (LVD function) of the battery bank
- energy savings—sleep mode turns inverter "off" but it comes back on at a flick of the first house light switch
- automatic start and stop of a backup generator
- maximum Power Point Tracking for grid-interconnected systems
- full safety protection for both homeowners and utility workers

An inverter installed for grid-interconnected operation is shown in schematic form in Figure 12-2. Energy from the renewable source is stored in the battery and directed to the inverter's internal components. A power supply and controller determine the sequence of events required to make the unit function as desired. When the inverter is activated, the controller starts a

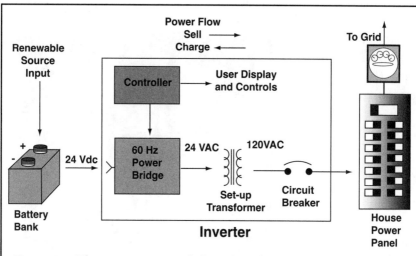

Figure 12-2. The inverter is a marvel of complex technology and is surprisingly easy to operate. A basic inverter for a grid-interconnected installation is shown in schematic form.

high-powered oscillator or power bridge that generates AC power at the voltage of the renewable source or battery bank, typically 12 V, 24 V, or 48 V.

The low-voltage AC power is set at a frequency of 60 cycles per second or 60 Hertz for the North American market. Inverters in Europe and Asia operate at 50 Hertz. You will recall from Chapter 1.3 that 60 Hertz means that the polarity of the voltage is switched back and forth 60 times per second. Imagine a simple C cell battery inserted into a flashlight first one way then the other 60 times per second, as shown in Figure 12-3. If we plotted the polarity of the battery each time it was inserted into and out of the battery clips shown in "A", we would obtain the plot shown in "B".

The advantage of AC power over direct current is that the voltage can easily be stepped up or down using a transformer. A transformer inside the inverter receives the low-voltage AC power and steps it up to the utility-standard 120 or 240 volts. In the inverter shown in Figure 12-2, the transformer is designed to accept a nominal 24 V AC input from the power bridge and convert it to a 120 V output. This power is then fed through a protective circuit breaker and into the power panel, which distributes it throughout the house. If the system is producing more energy than the house requires, the excess is automatically "exported" to the electrical utility via an electrical energy meter. Likewise, if the system is producing less energy than the house requires, the difference in energy demand is purchased or "imported" from the grid.

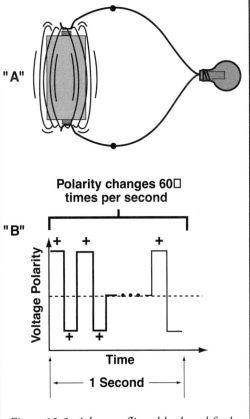

Figure 12-3. A battery flipped back and forth in its socket 60 times per second will supply alternating current to the light bulb load. The waveform at "B" shows the resulting plot of the changing polarity over a one-second interval.

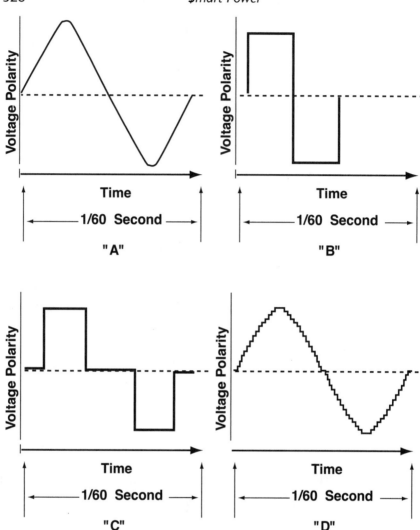

Figure 12-4

Inverter AC Waveforms

The electrical utility generates its AC power using mechanical generators connected to turbines. When a rotary generator creates an AC cycle, the waveform is sinusoidal, as shown in Figure 12-4 A. Compare this waveform with that of the simple inverter shown in Figure 12-3 and again in Figure 12-4 B. Each waveform has been magnified to show what is known as one cycle, one-sixtieth of a second long. Obviously, sixty of these cycles occur in one second.

The main difference between the two waveforms is that the utility-generated power shown in "A" has the voltage ramping up slowly until it reaches a peak level. It then falls back down slowly as the polarity reverses. In contrast, the simple inverter shown in "B" snaps quickly between points of opposite polarity, thereby acquiring the name "square wave." Only the least expensive and oldest inverters operate using the simple square wave.

Improvements in technology led to the "modified square wave" (also known as a "modified sine wave," but that's really pushing it), as shown in waveform "C". These inverters still retain a significant share of the market, but they are only used in off-grid and cottage applications and cannot be used in grid-interconnected systems.

Inverters using a square or modified square wave can operate 98% of all modern electrical appliances with no problem. However, the fast-switching edges of the waveform can produce a buzzing or humming noise in some items such as cheap stereos, ceiling fans, and record players. One primary advantage of this waveform is the ease of producing it. Inverters utilizing this pattern are robust, electrically efficient, and relatively inexpensive. They are not, however, compatible with the electrical utility system.

A fairly recent development in inverter technology is the sine wave model that outputs a waveform similar to that shown in "D." This digitally synthesized waveform shape is created using a technology similar to that used to record digital sound on a compact disc MP3 file, and we all know how good those sound. In fact, today's inverter technology offers negligible waveform distortion, with frequency and voltage tracking far better than that of even the most expensive utility power station. In the last few years the number of models and manufacturers has exploded, increasing the range of sizes and features while lowering prices.

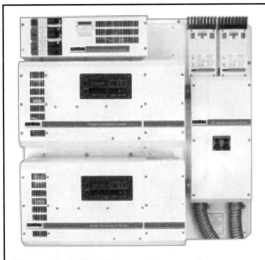

Figure 12-5. Two Xantrex sine wave inverters are "stacked" to provide both 120 V and 240 V AC power at better than utility-quality power. Also shown are two model C60 charge controllers (upper right) and a DC circuit breaker (bottom right). (Courtesy Xantrex Technology Inc.)

Grid-Interconnection Operation (grid-dependent mode)

A sine wave inverter that generates AC power of utility quality should be able to supply power to the grid in a manner similar to a hydroelectric turbine. Ten years ago this was considered heresy by the electrical utilities (and still is by some). Today, however, the demand for cleaner renewable energy (including political head-knocking by environmental and industry advocates) has weakened the monopoly previously enjoyed by utilities. Of course, some areas of North America are slower than others at getting the point, but the change is already occurring and will inevitably continue. Take California, for example: after continuous problems with rolling blackouts, sky-high energy costs, and never-ending smog, the politicians got the message. Now anyone

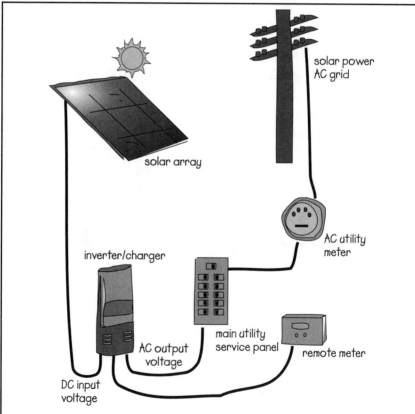

Figure 12-6. A typical grid-interconnected inverter is shown being fed by a PV array. This simple connection allows the electrical meter to spin backwards when energy is being exported (sold) to the grid. In cloudy times, you import (purchase) power from the grid as usual. The AC utility meter keeps track of these buy-and-sell transactions.

can install a bunch of PV panels on the roof, connect them all through a grid-interconnected inverter, and watch the electrical meter spin backwards. If that isn't enough, the state will even provide an incentive payment to help make it happen.

The grid-connected system shown in Figure 12-6 is typical of a simple, effective means of allowing renewable sources to supply energy to the utility. In this arrangement, electrical energy flows from the grid, through the utility meter, and into the power panel and electrical loads of the home. This is the normal *import* or purchasing mode that most homes are familiar with. During importation of power, the meter records the power usage on one electronic counter.

When the sun starts to shine on the PV array or the wind or water turbine starts spinning, the inverter synchronizes its own sine wave to that of the utility. In order to *export* power, the inverter creates a waveform that has a higher voltage than that of the utility source, causing current to be pushed onto the grid. During power exportation the utility meter records the generated energy on a second electronic counter. At the end of the billing period, the meter reader will record the power imported and exported to your house. The invoice you receive from the utility company will reflect the difference between the two meter readings, providing you with a net energy bill. (Refer to Chapter 7 for a discussion of the various types of revenue metering that improve upon the simple net metering program described here.)

For safety reasons, the inverter periodically checks the utility voltage to make sure that it is still present. If it is not, the inverter assumes the grid has failed. It then shuts down and waits until several minutes after grid power has returned to normal before re-applying power. This test eliminates the possibility of a condition known as *islanding*, which occurs if power from the grid is knocked out but the inverter continues to feed electricity into the local area. If this exporting of power to the grid occurs and a utility worker unsuspectingly touches the wires in the islanded area, serious injury can result.

Grid-Interconnection with Battery Backup (grid-interactive mode)

Figure 12-7 illustrates the essential components of a grid-interconnected system with battery backup, also known as a grid-interactive mode of operation. Using this arrangement it is possible to provide battery storage to supply power during periods of grid blackout. The major difference between grid-dependent and grid-interactive modes is the requirement for battery

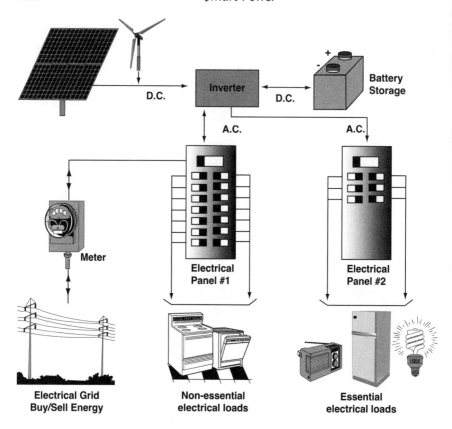

Figure 12-7. A grid-interactive system configuration provides electrical power to essential loads during grid blackout periods. This is accomplished by storing energy in a battery bank and wiring the home with two power panels, one for essential loads the other for non-essential ones.

storage and a secondary electrical supply panel to feed essential loads within the house.

When the utility supply is present, electricity is imported and exported in the same manner as in a grid-dependent system. During times of grid interruption, you will recall that the inverter in a grid-dependent system will disconnect itself and enter a standby state waiting for the grid connection to return to normal. With a grid-interactive configuration, the inverter will disconnect from the electrical utility and reconnect itself to an "essential electrical load supply panel." Using energy stored in a battery bank as well as from any renewable sources, the inverter will generate sufficient power to operate these connected loads. It is recommended that all the essential electrical loads be of the highest efficiency and lowest wattage, thereby prolonging battery life and corresponding electrical generation during a blackout.

Inverter Ratings

Important considerations when purchasing an inverter are the electrical ratings of the unit and whether or not you should purchase a "modified square wave" or "sine wave" model. Following is a list of items to look for when shopping for any inverter.

Sine Wave Versus Modified Square Wave Inverter Models

For grid-dependent or grid-interactive systems there is no choice but to use a *Grid Tie Sine Wave Inverter*.

For off-grid or emergency backup power systems that are *not* grid-interconnected you can choose between either sine wave or modified square wave inverter models. Modified square wave inverters tend to be less expensive than their sine wave counterparts and will operate almost all home appliances. The decision in this case will be based on budget as well as on the features you find desirable.

Output Voltage

North American homes are wired to the grid so that loads may be connected at 120 V or 240 V. Normal wall plugs are rated at 120 V, while heavy electrical loads such as electric stoves and furnaces, clothes dryers, and central air conditioning units operate at 240 V.

The voltage rating of most inverters is set to 120 V, allowing the operation of most household appliances and also making grid-interconnection possible. Increasing the size of the renewable energy source beyond the rating capacity of the inverter necessitates the addition of a second inverter. As a general rule of thumb, systems with a generating capacity of greater than 4 kW will require two inverters. When a second inverter is added, the electrical connection is in series, making the voltages additive at 240 volts.

In most cases the only 240 V essential loads that can be reliably connected are domestic water pumps. The vast majority of 240 V loads comprise appliances which draw enormous amounts of electrical energy, like central air conditioning, cook stoves, clothes dryers, swimming pools, and hot tubs. Trying to power these large loads during a grid failure while running on battery backup is the home-style equivalent of replicating the blackout of 2003. Battery capacity is limited and should be used sparingly. Most 240 volt loads should be wired to the inessential electrical panel.

If a 120 V inverter is specified for your application and you require 240

V supply to your essential load electrical panel, an auxiliary step-up transformer may be installed at considerably less cost than a second inverter.

Inverter Continuous Capacity

Inverter capacity refers to the amount of power that the unit can supply at
one time continuously. If the system is grid dependent, the inverter rating
should be at least 25% higher than the maximum power delivered by the
renewable sources. Remember that multiple inverters may be wired in series
and parallel configurations to increase the capacity as required.

If the system is grid interactive, the inverter capacity calculation will be
based on the same rules as for grid-dependent systems, as well as the maximum electrical load connected to the essential power panel.

You will recall that power, expressed in watts, is the voltage multiplied
by the current. When operating a home, you are likely to use more than one
electrical load at a time. In order to determine the inverter continuous capacity required during a blackout, it is necessary to total all of the essential
electrical loads that are likely to be turned on at any given time. For example,
assume that a washing machine (500 W), television and stereo (400 W),
refrigerator (400 W) and a bunch of CF lamps (100 W) are all on at the
same time. The total power requirement of these loads is:

$$500 + 400 + 400 + 100 \ W = \textit{Total Continuous Power}$$
$$= 1,400 \ W$$

This means that you will require an inverter capable of supplying at least
1,400 W continuously for the essential loads. If the grid-supplying power
rating is higher than the essential load power rating, the inverter capacity is
based on the former, with a 25% safety factor added. For example, if the
renewable sources power-generation rating is 2,500 W, the inverter capacity
should be at least 3,100 W.

2,500 W renewable generation x 125% capacity factor = 3,100 W inverter

If you are planning a grid-interactive configuration, remember to calculate the essential load value when filling out the *Electrical Energy Consumption Worksheet* in Appendix 7. When you are shopping for an inverter, you
will find that they tend to be sized in "building block" sizes. There are numerous small models below 1,000 W, although 2,500 and 4,000 W are common sizes.

When trying to read the continuous rating of an inverter, you may be
confronted with the term "VA" rather than watts. VA refers to the voltage

multiplied by the current that we have understood to mean wattage plus an allowance for power factor. At the risk of trying to split hairs, when an AC motor load is operated, there is an effect known as *power factor* that has to be taken into consideration. In the average home, VA closely approximates watts and may be used interchangeably. It is recommended to reduce the rating of the inverter by as much as 20% if you are operating air conditioning units, pool and spa pumps, or other similar high-wattage *induction* motor loads. If you are operating a shop full of large woodworking tools with many of the motor loads running at the same time, it is best to allow for an inverter derating of approximately 25% or higher to allow for power factor. In practice there is little that you can do other than to be aware of the existence of power factor and reduce simultaneous usage of motor-driven appliances or purchase the next-larger inverter. An additional point is that universal motors that have brushes (sparks may be seen when the motor is running) can be run safely without any derating concerns as they are not affected by power factor issues.

Examples of universal motors include:
- regular and central vacuum cleaners
- food processors and mixers
- drills, routers, shopvacs, radial arm and circular saws
- electric chain saws and hedge trimmers
- electric lawn mowers

If in doubt, contact an inverter dealer or motor repair shop for clarification.

Inverter Surge Capacity

The surge capacity is an indication of how much short-term overload the inverter will be able to handle before it "trips" on this condition. Surge capacity is necessary to allow some large loads to get started, particularly motorized loads requiring starting power two to three times their running power. Although this start period is very brief and lasts a fraction of a second, it should be considered. The main concern is whether or not you have any "unusual" electrical devices such as an arc welder in your home. In addition, it is wise to look at your electrical appliance consumption list and see if there is a likelihood that several large motor loads may start at the same time while operating in battery backup mode. However, in the average home it is highly unlikely that surge capacity will become an issue.

Inverter Temperature Derating

The power protection circuitry for most inverters is temperature compensated, meaning that the maximum load that an inverter can run changes with the ambient temperature. As the temperature of the internal electronics of the power switching bridge increases, the allowable connected load current/power is reduced.

The graph in Figure 12-8 shows the effect of temperature on the capacity of a Xantrex model SW series 4,000 watt inverter to operate connected loads. Notice that the inverter reduces its capacity at temperatures above 77° F (25 °C). The graph also assumes that the inverter is operating at sea level and without any restriction of the airflow around it.

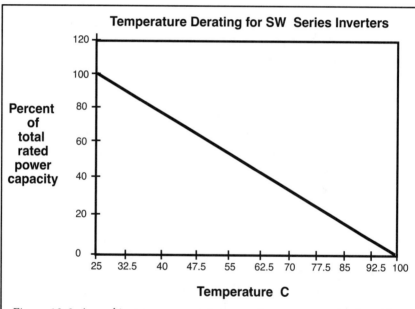

Figure 12-8 As ambient temperature increases, inverters must be derated as illustrated here or according to the manufacturer's data sheet. (Courtesy Xantrex Technology Inc.)

If your grid-interconnected inverter is designed for outdoor installation, it is advisable not to mount it in direct sunlight if at all possible. There have been numerous instances of inverters entering a self-protection mode while basking in the hot Arizona sun.

Battery Charging

You can save money by purchasing an inverter without a battery charger installed rather than a combined inverter/charger unit. If you want to install a battery charger after the inverter is installed, you can purchase a separate battery charger such as the TrueCharge™ series from Xantrex Technology Inc.

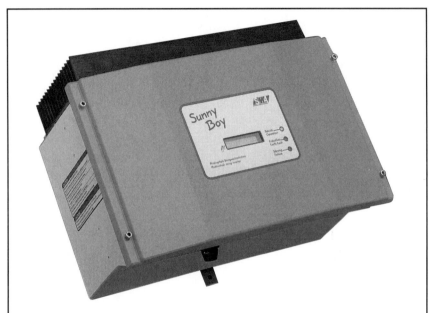

Figure 12-9. The SMA model Sunny Boy 2500 is an example of a grid-dependent inverter supplied without battery charging and other optional features.

13
Fossil-Fueled Backup Power Sources

If there is an Achilles' heel with home-based energy systems, it is not re-lated to renewable generation equipment reliability, but rather to the grid. The disastrous blackout in 2003 which plunged nearly fifty million people into total darkness has made a growing number of people increasingly aware of just how reliant we have become on a continuous, uninterrupted supply of electrical power. The prognosis for the future is not as rosy as many would like, and they are taking mat-ters into their own hands. Unfortunately, a grid dependent elec-trical generation sys-tem will not provide electricity when the utility supply goes down. Many folks cannot afford a grid interactive system to keep the lights on. What to do?

Figure 13-1. Fossil-fueled backup generators are the antithesis of what renewable energy stands for. Nevertheless, they are an integral part of any off-grid system. (Courtesy Generac Power System Inc.)

Bring in the fossil-fueled backup generator, the antithesis of what renewable and clean energy technologies are all about. Life is always full of tradeoffs, and the emergency backup generator is one that can't be ignored.

A backup generator is designed to perform one primary function: supply essential power to the home during a blackout and then shut down as quickly as possible. Generator power is polluting, noisy, and expensive compared with grid or renewable-source electricity. The less time the generator runs, the better.

Of course it is not necessary to use a fossil-fueled generator at all. A grid-interactive renewable energy system with photovoltaic panels and battery backup should be able to get most houses through the typical grid blackout. The operative word is *typical*; many people were astounded at the extent and length of the 2003 blackout. Grid-interactive renewable energy systems are normally sized to provide power to essential loads for approximately 12 to 24 hours on battery power alone. Although photovoltaic panels or a wind turbine may extend this time, if the weather happens to be uncooperative coincident with the blackout your batteries may find themselves depleted along with the grid. This is when the backup fossil-fueled generator can come to the rescue.

Generator Types

A backup generator with a reciprocating internal combustion engine is more correctly known as a *genset* (a *gen*erator and motor *set*). There are many shapes and sizes and a variety of fuel supply choices. Before we look at models suitable for emergency systems let's review what types not to buy.

Small generators such as the ultra-portable model shown in Figure 13-2 are not suited to full home-emergency or battery-charging applications. As a quick rule of thumb, if you can lift the generator, it is probably too small. The Honda

Figure 13-2. This small 1,000 W ultra-portable unit from Yamaha is fabulous for operating small power tools or emergency lights, but it is not suited for a house full of appliances or for battery charging. (www.yamaha.com)

model EM5000S shown in Figure 13-3 is about the smallest (and least expensive) generator recommended for this application. If you have an old 4,000 W unit kicking around in the garage, by all means put it to use. Keep in mind that performance, fuel economy, and battery charging time will be compromised with undersized models.

Generator Rating

Generators, like inverters, are rated in watts (W) or more correctly volt-amps (VA). The reason for this is that when motor loads are connected to an electrical circuit they behave differently from resistive loads such as lights and heating units. Although the generator may have sufficient nameplate rating capacity, certain loads might not run because of the effects of power factor, necessitating the distinction between VA and W .

Figure 13-3. The Honda EM5000S or equivalent-sized models are the smallest units that should be considered for emergency backup power.

Inexpensive gensets will have inexpensive generators and support electronics. This is not a problem for many applications until big motor-driven devices are activated. Anyone familiar with gensets will know that air compressors, well pumps, and furnaces may have a hard time getting started even if the genset has a higher power rating than the connected load. Motor loads have high starting power requirements that can exceed their normal requirements by three or four times.

Similarly, battery charging can be very hard on smaller, less expensive gensets because a battery charger consumes power only from the very peak of the alternating current waveform. Smaller units, typically less than 5,000 W, have difficulty providing power in this mode, even though the charging power applied to the battery may be a fraction of the generator's rating.

A gas engine driving a generator is a reciprocating device. You will recall from earlier chapters that rotating generators produce an alternating current (AC) voltage which traces a sine wave pattern. An example of this wave form is shown in Figure 13-4, with voltage represented on the "y" axis and time on the horizontal "x" axis. Starting from the extreme left of the waveform, the voltage starts out with zero amplitude and slowly rises until it reaches a peak of 170, at which time the polarity starts to reverse and the voltage drops back to zero before starting on the negative half of the cycle.

You may be wondering why the genset output voltage is 170 as opposed to the 120 volt level we are accustomed to. Because AC sine wave voltages change with time, we are faced with the issue of determining where on the waveform the voltage measurement should take place. A mathematical formula called the "root mean square" can be applied to such a waveform to calculate the voltage present in the "area under the curve". Perhaps a simpler, less technical way of visualizing the voltage of a sinewave is to refer to it as the average level. If you were to take instantaneous voltage measurements of the sine waveform you would find that it rises from zero and peeks at approximately 170 volts, as discussed above. It is the instantaneous voltage level that provides battery-charging capability.

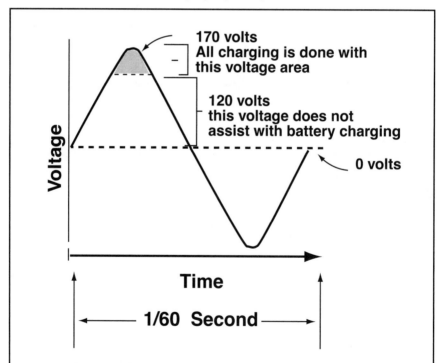

Figure 13-4. Battery-charging applications use only the top portion of the generator's sine wave voltage output. For this reason small generators lack the output voltage required to charge batteries.

The generator's peak output voltage must be high enough for battery charging to occur. The battery-charging unit inside an inverter contains a transformer that is capable of stepping up or down the applied voltages depending on which mode of operation is selected. For example, with a 12 V battery connected, the inverter can step this voltage up to 120 V AC to operate connected loads. To charge a battery, the inverter can reverse this process by stepping the generator's 120 V AC down to a DC voltage level sufficient to "push" current into the battery under charge. The waveform in Figure 13-4 shows that the generator's AC voltage starts at zero and climbs to a peak of 170. There will be a period where the generator's "stepped-down" voltage is less than that of the battery bank. No charging current will flow at this time.

Once the generator's *instantaneous* voltage exceeds 120 V, the inverter's transformer will step it down by the appropriate ratio, convert the AC to DC, and feed the voltage into the battery. The example shown in Figure 13-5 will help to clarify this point. The generator instantaneous voltage has risen to 150 V. The applied voltage reaches the inverter's battery charging transformer where it is stepped down to the ratio determined by the nominal battery bank voltage, which is 10:1 in this example. The transformer output voltage is 15 V AC. This voltage is applied to a rectifier that converts the alternating current to direct current. As the 15 Vdc output is greater than the battery's 12.18 V (or 50% discharged) reading, current will flow into the battery.

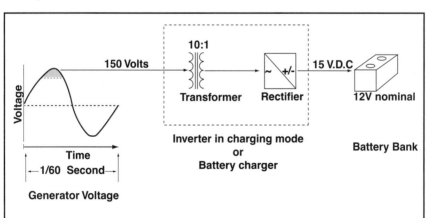

Figure 13-5. It takes a strong, high-quality generator to support battery charging and operate large motor loads such as furnaces and well pumps. Weak or inexpensive generators burn excessive amounts of fuel during the long runtime required to charge a battery.

The "squashing" of the peak of the sine wave identifies gensets with weak or small generators that have insufficient power to support battery charging. Longer charging time and increased fuel consumption result in unnecessary genset wear and a higher cost for each watt of electricity stored in the battery.

It is highly recommend that you purchase a genset with a rating of 7,500 W (120 V/62 A or 240 V/31 A) or higher. Check with the manufacturer to determine if the unit is equipped with an electronic voltage regulator module within the generator. High-quality generators often have a peak voltage adjustment that ensures rapid battery charging and immediate starting of difficult motor loads. Discuss this issue with the generator sales staff. If they are not familiar with battery charging applications for a particular model of generator, provide them with a copy of the above text and ask them to review the issue with the factory. It is pretty tough to return a generator that is too small for your application.

Generator Type	Inverter Type	Typical Maximum Charging Current (Amps)
Honda 800	DR1512	43
Honda 2200	DR1512	57
Homelite 2500	DR1512	11
Honda 3500	DR1512	39
Westerbeke 12.5 kW	DR1512	65

Table 13-1. This table produced by Xantrex Technology Inc. compares the charging current of several generator models when using a 12 V inverter/charger. The Westerbeke 12.5 kW model can charge a battery bank six times faster than a Homelite 2500. This equates to reduced generator running time, wear, and fuel consumption.

Voltage Selection

The voltage selection of the genset will be determined by the connection method with the home and whether or not essential loads require 240 volts. It will be necessary to review voltage selection with your electrician during the planning phase.

See Chapter 14, "Putting It All Together Safely", for more information on generator voltage selection and connection.

Fuel Type and Economy

Gensets are available in several fuel choices including gasoline, natural gas, propane, and diesel. The less expensive units tend to be equipped with gasoline engines. Larger industrial-grade models are normally fueled using propane or diesel. The choice depends on several factors such as cost of the unit, proximity to a fuel source, and desire for economy and ease of use. From an environmental perspective, look for engines that have an EPA (Environmental Protection Agency) rating, are four stroke, are slow speed (typically 1800 rpm or slower), and preferably burn natural gas. Fuel types that you can choose from are as follows:

- **Gasoline:** Everyone is familiar with the small gasoline engines that are ubiquitous throughout North America. Cheap and easily fueled, they have a very short life span when used in demanding applications. The majority of gasoline engines operate at 3,600 revolutions per minute (RPM), which results in rapid wear and high noise levels. Expect a life span of five years before a major rebuild is required.

- **Natural Gas:** Natural gas engines are offered in two varieties: converted gasoline and full-size industrial. The converted engine is really no better than a gasoline engine, except that it offers the advantage of no fuel handling because it can be directly connected to the gas supply line. Industrial-sized natural gas engines are of a heavier design and operate more slowly, typically at 1800 RPM. This increases engine life and greatly reduces engine noise. However, natural gas is not available in all areas because the fuel is transported via pipelines.

- **Propane:** Propane is similar to natural gas with the exception that this is the fuel of choice for off-grid and rural applications; propane may already be the fuel source for other appliances in your home.

- **Diesel:** The diesel engine has the best track record for longevity. Diesel units are heavy, long-lasting machines that operate at slow speeds. Fuel economy is highest with a diesel engine. Besides offering superior fuel economy, modern diesel engines have excellent cold-weather starting capabilities and are clean burning.

- **Biodiesel:** As the name implies, biodiesel is a clean burning, alternative fuel produced from domestic, renewable resources. Biodiesel contains no petroleum, but can be blended at any level with petroleum diesel to create a biodiesel blend (most often a blend called B20 with a ratio of 80% petroleum diesel to 20% biodiesel). It can be used in diesel engines with no major (or any) modifications. Biodiesel is simple to use, biodegradable, non-toxic, and essentially free of sulfur and aromatics. (See further notes concerning biodiesel fuel in Chapter 5).

A typical 8-kilowatt propane genset consumes 1.93 gallons (7.3 liters) per hour when operated at 100% capacity. An equivalent genset in a natural gas-fueled model will consume 144 cubic feet per hour (4,077 lph). An equivalent diesel model from China Diesel Imports (www.chinadiesel.com) requires only 0.78 gallons per hour (3 lph). A diesel (or biodiesel) model requires 40% *less* fuel than a propane model for the same amount of appliance operation or battery charging. Over the considerable life span of either model, this translates into a significant savings in operating cost. The fuel economy of a gasoline engine is comparatively poor.

Another argument against gasoline is the requirement to pay "road tax" when you fill up at the local gas pump. Less expensive colored or "off road" gasoline can be purchased but may be difficult to locate. Even with the reduction in road taxes, a high-speed (3,600 RPM) gasoline model will not be as economical (or as quiet) as a diesel or low-speed natural gas or propane model.

Generator Noise and Heat

It's annoying enough to have to run a genset in the first place, but it's even more aggravating to have to listen to it running. The best way to eliminate this problem is not to operate one at all. The next best plan is to locate the unit a reasonable distance from the house and enclose it in a noise-reducing shed or chassis.

The Kohler natural gas genset shown in Figure 13-6 and the equivalent Generac model shown in Figure 13-1 are mounted in a noise-deadening, weatherproof chassis which may be mounted on a cement pad in a similar manner to central air conditioning units.

You can either build a noise-reducing shed using common building materials or purchase a wood-framed toolshed building. The shed should be fabricated with a floating deck floor that does not contact the walls of the building. This construction prevents engine noise and vibration from radiating outside the building. The walls should be packed solidly with rock wool, fiberglass or, best of all, cellulose insulation. The insulation should then be covered with plywood or other finishing material, further deadening sound levels.

All internal combustion engines create an enormous amount of waste heat. A little dryer vent or hole in the wall won't cut it. The unit shown in

Figure 13-6. This Kohler natural gas genset is manufactured with a noise-deadening, waterproof chassis and provided with fully automatic controls. It will automatically start and provide power to the house as soon as the electrical grid fails. Once utility power returns, the unit will reconnect the house to the grid and shut itself down. Units can even be programmed to automatically exercise themselves to start and stop periodically, ensuring they will be ready at the next blackout.

Figure 13-7 is mounted so that the 18" x 18" (0.5m x 0.5 m) radiator and fan assembly blows outside the building pointing away from the main house. This arrangement also requires an air intake, which is provided by air passing under the floating deck of the building.

The exhaust gas leaves the muffler vertically and passes overhead to a second automobile-style muffler before exiting the building, further reducing noise.

Generator Operation

If you want to move the genset away from the house, there are two drawbacks you need to consider. A suitable power feed cable of sufficient capacity will have to be run either overhead or in an underground trench, and long cable runs increase the cost of installation. The second consideration involves starting the genset. If the unit is equipped for manual starting, it will be necessary to go to the machine shed each time you wish to start and stop the unit. This is no problem on a nice summer day, but it can become a bit trying during winter storms when you need the darn thing the most.

Gensets equipped for automatic operation as discussed above solve the starting problem completely. Alternatively, when the AC supply cable is placed

Figure 13-7. Installing a genset in a well-insulated shed with a "floating" floor prevents motor noise and vibration from radiating outside the building. Include a large vent fan to remove waste heat, and ensure that it is directed away from the main house.

in a trench, a second "control" cable can be run alongside it. This cable is connected to the genset's control unit and the other end is connected to a manual start/stop switch inside the house or to the automatic generator controls contained within an inverter.

Automatic controls do not add an appreciable amount to the cost of the genset and will greatly improve your relationship with the beast.

Other Considerations

A genset should have a long life, particularly if the unit is well maintained. Many underestimate how much television they watch, and I suspect that estimations of generator running time are a close second. Running time is not a problem, but knowing when to perform periodic maintenance on the unit is. For example, the Lister-Petter diesel engine shown in Figure 13-7 requires various servicing functions at 125, 250, 500 and 1,000-hour intervals. Order the unit with a running time meter at a small extra cost.

Oil, air, and fuel filters must be changed periodically. Order a service manual and sufficient spare parts for your model. The dealer will be able to recommend a suggested spare parts list.

Have engine oil on hand as well as the necessary tools to change filters and drain the engine crankcase. Inquire at your local garage about where used engine oil can be dropped off for recycling. Most garages will be happy to oblige, especially if you deal with them for automotive service. Never pour used motor oil into the ground or down a storm sewer. One quart (one liter) of oil can easily destroy 100,000 quarts of ground water, seriously damaging the environment.

To ease winter starting, use synthetic oil rated for operation with your generator model.

Further Notes On Biodiesel Fuel

In recent years, there have been remarkable strides in the "greening" of diesel engines, to the extent that many are compliant with California's Clean Air Act. Anyone who thinks that diesel engines are slow, clunky, and smelly have obviously not been introduced to the new "common rail diesel engine" technologies of the past few years. Witness the Mercedes Benz E320 series. This car is just as quiet as its gasoline counterpart and faster in both acceleration and top speed. Cold weather starting problems are also a thing of the past. Not only can you forget about block heaters, but the whole glow plug thing has gone the way of carburetors, muscle cars, and fuzzy dice. (OK, maybe not the fuzzy dice.)

As a further advantage, biodiesel fuel (manufactured using renewable soya and other grains as well as waste grease and animal fats) may be available in your area. Burning biodiesel is considered green since it has lower life-cycle carbon dioxide and smog-producing emissions than its fossil-fuel counterpart. There may be a slight cost penalty for its use, but considering the minimal running time of the genset and the symbiotic relationship this fuel offers with the renewable energy system, it is well worth the price. Biodiesel might even improve your relationship with the genset.

You will find more information on biodiesel in Chapter 5.

14
Putting It All
Together Safely

We have finally made it to the point where we can stop talking about how all the bits and pieces that make up a renewable energy electrical system work and start putting it all together; *safely*. This chapter deals with interconnecting the various system components in a neat and effective arrangement.

If you are not familiar with electrical wiring, conduit, and general construction work, it's still well worth having a look at this section in order to understand what your electrician is talking about *before* you "throw the switch." This chapter and the relative appendices can also act as a reference should your electrician not be familiar with some of the details of working with direct current (DC), PV modules, and wind turbines.

WARNING!

Grid-interconnected systems are a fairly new phenomenon in North America that may not be allowed in your jurisdiction. Be sure to check with your electrical inspection authority before you commit to such a system. Also, be aware that even where connection is allowed you may not be paid for the electricity you produce.

A Word or Two about Safety

Obviously, you want the installation work to be done correctly and safely. Owning and operating a renewable energy system is quite enjoyable, and you can almost forget that you have one at times. Although it is pretty cool stuff, it is not a toy and can cause electrocution or fire hazards if not respected. My first discussion with electrical contractors and inspection people left me bewildered. I clearly remember one person saying eleven years ago: "Why would you want one of those systems? You won't be able to run a toaster." Although I still chuckle at this while eating my morning toast, it goes to show that not everyone is up to speed with the technology.

Owning a renewable energy system is no different from owning and operating a standard electrical power station. Size doesn't matter. You can be killed or be seriously injured with battery or inverter power, just as you can with energy from the grid.

Electrical Codes and Regulatory Issues

In North America, electrical installation work is authorized by local electrical safety inspection offices that issue work permits and review the work in accordance with national standards. In the United States, the National Electrical Code (NEC) has been developed over the last century to include almost all aspects of electrical wiring, PV, battery, and wind turbine installation. In Canada, the Canadian Electrical Code (CEC) performs the same function as the NEC.

The NEC and CEC comprise the Part 1 Installation Codes which regulate the interconnection and distribution of electricity to industrial, commercial, and residential buildings. These codes also deal directly with the internal wiring of your home.

Many people believe that because they have their own renewable energy system the code rules do not apply to them. This is wrong. With few exceptions, the installation of PV and wind systems must comply with the requirements of the code. In fact, way back in 1984 Article 690 was added to the NEC to deal specifically with the installation of PV systems.

In addition to the CEC/NEC rules, a Part 2 product standard is required to certify every electrical appliance that operates at 120/240 volts (Vac). Where safety concerns exist, this standard may be extended to lower voltage products such as battery-operated power tools. Many people are familiar with the Canadian Standards Association in Canada and Underwriters Laboratories in the United States. Working in conjunction with the CEC/NEC, these safety agencies are charged with the development of electrical and fire safety standards for household appliances. When a manufacturer develops a new product, the design must undergo extensive safety-related tests by these agencies. Products that meet the requirements are eligible to carry a "certification mark" which tells electrical inspectors that when properly installed they will be safe.

Legitimate manufacturers have their products undergo such testing and are eligible to use the UL, CSA, ETL or other authorized testing laboratory seal of approval. When comparing and purchasing products, look for this seal as a sign of a safe, quality design and be aware that electrical devices without it will not be allowed to connect to the grid.

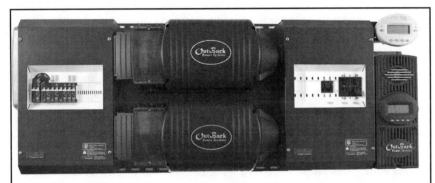

Figure 14-1. Renewable energy electrical installation is not difficult, as this prewired, integrated panel from Outback Power System shows. Viewing the panel from left to right you see the AC circuit breakers, 2 x 120 Volt inverters "stacked" for 240 Volt output, DC circuit breakers for PV panel, and optional battery bank and MPPT series voltage regulator.

When you or your electrician is ready to begin wiring, it will be necessary to apply for an electrical permit. This permit will authorize you to:

- Perform all electrical wiring according to NEC/CEC codes and any local ordinances in effect at the time of installation.
- Install only electrical equipment that is properly certified. Each device must have a UL, CSA or other approval agency certification "mark."
- Provide the inspector with copies of wiring plans, proof of certification, or other engineering or technical documentation to aid in his or her understanding of the renewable energy system.

Give the inspector written notification that the work is ready for inspection. You must not cover or hide any part of the wiring work, including backfilling of trenches, until the inspector has completed the inspection and provided an authorization certificate.

Your electrical inspector is not working against you. If he or she is asking a lot of questions it is to understand what you are doing. Renewable energy systems are not yet considered mainstream technology, and some inspectors may not be familiar with the specifics. On the other hand, your inspector will know an awful lot about wiring and installation details, and most professionals will be more than happy to assist with guidance and pointers.

Electrical inspectors will review the design and installation work and check for certification marks on the various appliances. Some system components on the market may not have test certification markings, but all electrical code rules require that products *must* have them. It is highly recommended that you check any products before you purchase them to ensure proper compliance. If you require a product and the manufacturer has not had it tested, discuss this with your inspector before you buy. A certified product may be available, albeit at a higher cost. Alternatively, field inspection on site may be allowed in your jurisdiction for an additional fee.

CAUTION!
As code rules are updated on a regular basis and may have subtle differences from one locale to another, use the information in this chapter as a guide but discuss the details with your electrician and inspector before proceeding with installation work.

What Goes Where?

In previous chapters we have dealt with each component as a separate piece of the pie. We are now ready to begin planning the wiring installation to interconnect all of the components.

The wiring overview in Figure 14-2 illustrates how interconnections are made between each component of the *grid-dependent* system. PV arrays always generate direct current (DC) and may be connected directly to the inverter for conversion to alternating current. Wind turbines may operate using direct or alternating current (AC), although in the case of the latter configuration a unit known as a *rectifier bridge* will convert the voltage to DC for supplying the inverter.

The configuration for a *grid-interactive* system which provides emergency backup power during blackout periods is shown in Figure 14-3. The feed

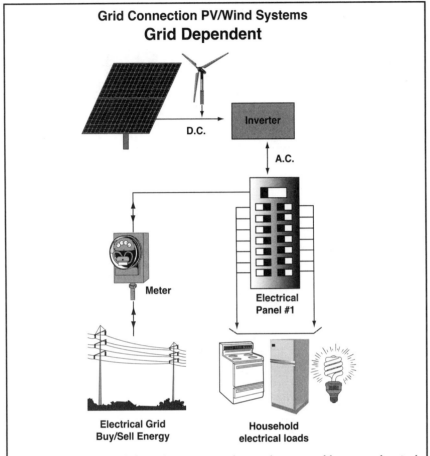

Grid Connection PV/Wind Systems
Grid Dependent

D.C.

Inverter

A.C.

Meter

Electrical
Panel #1

Electrical Grid
Buy/Sell Energy

Household
electrical loads

Figure 14-2. The grid-dependent system is the simplest renewable energy electrical configuration. It will generate power for your home and automatically sell any surplus to your electrical utility. With this installation there is virtually no maintenance required; just sit back and collect the money.

from the PV panel or wind turbine, battery bank, energy meter, charge control, diversion load, and inverter circuits will all be completed using direct current.

In either system the AC connections between the inverter and house supply panel follow standard household electrical wiring.

The distinction between DC and AC wiring is very profound. Wire, connectors, fuses, and switches are generally not interchangeable. Because the current on the DC side is very high (owing to the lower voltage), wire size tends to be much larger than on the AC side. For this reason we will review each wiring component separately.

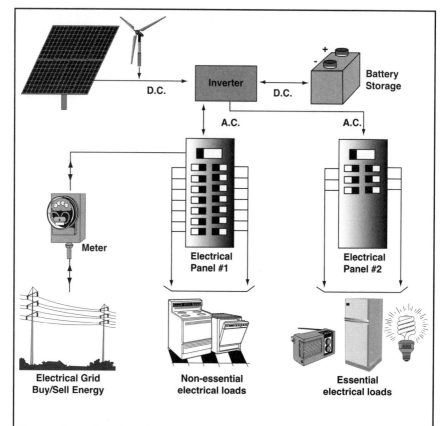

Figure 14-3. When the utility is operating, the grid-interactive configuration works in the same manner as the grid-dependent design. During a blackout, energy is supplied to essential house loads either from the renewable energy source or from a battery bank.

Direct Current (DC) Wiring Overview

Energy stored in a battery bank or supplied by a PV panel is typically supplied at a low DC voltage. When energy is required for our homes, the DC voltage is converted to AC by the inverter and stepped up to either 120 V or 240 V. If we assume that the inverter will supply a maximum house load of 1,500 watts (W) we know that 1,500 W plus an allowance for inefficiencies has to flow out of the battery bank. The wattage, voltage, and current relationship for the household side of the inverter is:

1,500 W load ÷ 120 V house supply = 12.5 amps (A) current flow

And on the low voltage input side of the inverter:

1,500 W load ÷ 12 V battery supply = 125 A current flow

A quick rule of thumb is to remember that the low-voltage side of the inverter current is 10 times greater than the 120 Vac side when dealing with 12 Vdc configurations, 5 times greater for 24 Vdc, or 2.5 times greater for 48 Vdc.[1]

It stands to reason that, since the wattage is the product of the voltage and current, lowering the voltage will cause a corresponding rise in current and vice versa. It requires a wire the size of a Polish sausage to carry 125 A of current in a 12 V circuit yielding 1,500 W. The same 1,500 W can be supplied through a light-duty extension cord if the voltage is cranked up to 120 V.

Large-gauge copper wire is expensive and difficult to work with, offering us plenty of reason to stick with higher DC voltages where possible. This is also the reason that large AC energy consumers in the home such as electric stoves, furnaces, central air conditioning, and dryers are always rated 240 V, allowing the use of smaller wire sizes.

Figure 14-4. Direct-current wiring circuits such as the battery interconnection cables shown here are expensive and difficult to work with. Keep wiring runs as short as possible and purchase premanufactured cable such as this whenever possible.

---

[1] Grid-dependent inverters such as the SMA Sunny Boy model (Figure 12-9) accept high-voltage directly from PV panels. Wiring a sufficient number of PV panels in series raises the operating voltage on the DC side, eliminating the need for a large power transformer inside the inverter while reducing the wire size required for installation.

Low-voltage electricity is also difficult to transmit any significant distance. As we increase the current flow to compensate for the lower voltage, more of the power is because of the resistance of the wire. The only way to address this problem is to raise the voltage, decrease the transmission distance, or increase the size of the wire. Often it is necessary to do all three. In short, keep DC wiring runs (between PV panels and the inverter or battery) as short as possible and increase the voltage where practical. Appendix 9 contains tables that relate the system voltage and load current to wire size and provide a maximum one-way cable length. This chart assumes a 1% electrical voltage loss, which in turn causes a corresponding loss in power. If you can tolerate a higher level of loss, then each of the applicable distances and losses may be doubled, tripled, etc.

For example, in the table for 24 V systems in Appendix 9, assume that a PV array is delivering 40 A of current and requires a cable run of 24 ft (7.3 m). The chart indicates that a #0 size of wire (some trades refer to this as "#1/0", while "#00" is referred to as "#2/0") will be required to maintain 1% electrical loss. Wire of this size is pretty big (about the diameter of a pencil), hard to work with, and fairly costly. Let's consider some alternatives:

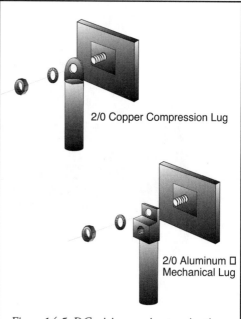

2/0 Copper Compression Lug

2/0 Aluminum ☐ Mechanical Lug

Figure 14-5. DC wiring requires terminations that are clean and electrically reliable. The copper compression lug on the left is a premanufactured cable for high-powered inverter and battery connections. The aluminum lug on the right can be assembled on-site without the need for special tools.

1. Do nothing. Use the #0 wire and call it a day. There is nothing really *wrong* with using #0 wire; bigger wire is simply more expensive, harder to work with, and difficult to interconnect.

2. Increase the system voltage to 48V, thereby decreasing the cable size to #6 gauge. (Remember, doubling voltage halves the current to 20 amps.)

3. Allow an increase in voltage loss. Leave the system voltage at 24 V, but allow the system voltage drop to increase to 4%. Use the 24 V data table and reduce wire size as desired. For example, using a #6 wire will increase losses to 4%. (#6 wire will carry 40 amps with 1% loss 6 feet. Increasing this length to 12 and 24 feet will increase losses to 2% and 4% respectively).

4. Check the table in Appendix 10 to be sure the selected wire size can carry the required current. Number 6 wire is rated for a maximum of 75 amps. Don't go overboard with losses to increase wire-run distances, as losses will reduce valuable power supplied to the inverter.

5. Run two sets of smaller wire. Suppose that we are connecting a PV array to a battery bank. It is possible to "split" the wiring of the array in two, reducing the current in each set to 20 A. If each set of arrays is connected with #4 wire, we can still maintain our voltage drop at 1% and use smaller wire, which is easier to handle. The two arrays would be connected in parallel back to the inverter or battery bank, combining to produce our 40 A supply.

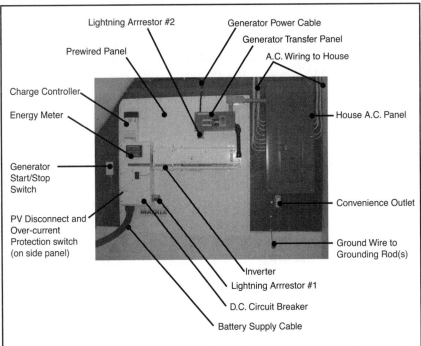

Figure 14-6. Renewable energy electrical installation is quite easy, as this pre-wired off-grid integrated panel from Xantrex Technology shows. The large gray panel to the right is where the standard 120 V house wiring begins.

The low-voltage side of an inverter requires an enormous amount of current, so the larger the cables connected to the inverter the better. Undersized cables result in additional stress on the inverter, lower efficiency, reduced surge power (required to start motor-operated loads when using emergency battery backup power), and lower peak output voltage. Don't use cables that are too small and degrade the efficiency that you have worked so hard to achieve.

In addition, keep the cable runs as short as possible. If necessary, rearrange your electrical panel to reduce the distance between the batteries and the inverter. The lower the DC system voltage, the shorter the cable run allowed. If long cables are required, either oversize them substantially or switch to a higher system voltage as discussed above.

Although large cable may seem expensive, spending an additional few dollars to ensure proper performance of your system will be well worth the investment. For their inverter systems, Xantrex Technology recommends that the positive and negative wires be taped together to form a parallel set of leads. This reduces the inductance of the wires, resulting in better inverter performance.

Cable Size Required	Rating in Conduit	Maximum Breaker Size	Wire Rating in Air	Maximum Fuse Size
#2 AWG	115 amps	125 amps	170 amps	175 amps
00 AWG	175 amps	175 amps	265 amps	300 amps
0000 AWG	250 amps	250 amps	360 amps	400 amps

Table 14-1. Battery and inverter cable sizing chart. Note that as a result of heat loss, wires run inside a conduit have a much lower rating than those exposed to air. Battery-to-inverter cable runs should not exceed ten feet in one direction. It is recommended that you use #0000 AWG-size wire for all inverter runs, regardless of length.

AC/DC Disconnection and Over-Current Protection

For safety reasons, and to comply with local and national electrical codes, it is necessary to provide over-current protection and a disconnection means for all sources of voltage in the ungrounded conductor. This includes the connection between the PV array, wind turbine, and batteries as well as the connection between the batteries and inverter. Most AC sources are already provided with a protection and disconnection device as an integral part of the house or appliance wiring. Generator and inverter units are generally provided with their own internal certified fuse or circuit breaker device.

Standard AC-rated circuit breakers and fuses will not work with DC circuits, and such an installation should never be attempted. Fuses and circuit breakers such as those shown in Figure 14-7 may be used, as they are rated for breaking a direct current electrical source.

Fuses and circuit breakers are similar to safety valves. When the flow of current through a conductor or appliance exceeds its specified rating in amps, a fuse which is wired in series will "blow," opening the electrical path. A circuit breaker works in the same manner, except that it may be used as a temporary servicing switch and can also be reset after a trip condition.

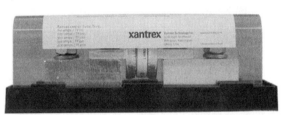

Figure 14-7. A "T" series DC fuse is shown above. On the left is a direct current disconnection and over-current protection circuit breaker and mounting box. (Courtesy Xantrex Technology Inc.)

Each circuit path will have a maximum current based on worst-case conditions. This current level must be calculated by reading the manufacturer's data sheets for the appliance, wire, inverter, PV panel, or other device. The current rating should be given a safety factor of 25%. Therefore, when sizing a cable for a run between a PV array and inverter, the cable-run distance, system voltage, and worst-case current and wire current-carrying capacity have to be considered. In addition, if the cable is contained in conduit that is exposed to the summer sun, the insulation temperature rating must also be considered. For example:

1. A PV array outputs a maximum of 30 A under all conditions and the one-way wiring distance is 40 ft (12 m). The system voltage is 24 Vdc.
2. From the chart in Appendix 9, we see that a #00 (#2/0) wire is required to carry this current with a maximum loss of 1%.
3. From the chart in Appendix 10, we see that a #00 wire is capable of carrying a maximum current of 195 A.

4. A safety factor of 25% is added to our maximum PV array current (30 A x 1.25 = 37.5 A).
5. The worst-case current calculated in #4 above is compared to the maximum current rating of the wire calculated in #3. If the worst-case current is less than the desired wire-size capacity, it is acceptable.
6. The circuit breaker or fuse size must be equal to or less than the maximum rating of the wire capacity as defined in the NEC/CEC. Note that electrical cables run in conduits or raceways must be derated because of heating effects.
7. The ambient temperature of the environment and other conditions affecting wire insulation temperature ratings are discussed in Chapter 2 of the NEC and Chapter 12 of the CEC.

These notes show you the complexity of wire selection, placement, and type. If you are not familiar with the above issues, purchase a copy of the latest NEC/CEC and review the appropriate standards. Alternatively, you can discuss these items with your electrical inspector at the planning stages.

To purchase the NEC contact:
National Fire Protection Association
1 Batterymarch Park,
Quincy, Massachusetts
USA 02169-7471
www.nfpa.org

To purchase the CEC contact:
Canadian Standards Association
5060 Spectrum Way
Mississauga, Ontario
L4W 5N6
ww.csa.ca

Battery Cables

According to the NEC, battery cables must not be made with arc welding wire or other non-approved wire types. Standard building-grade wire must be used. The CEC has no such restriction.

Wiring Color Codes

Wiring color codes are an important part of keeping the interconnection circuits straight when installing, troubleshooting, or upgrading the system at a later date. The standard color schemes used are discussed below.

Bare copper, green, or green with a yellow stripe

Used to bond exposed bare metal in PV modules, frames, inverter chassis, control cabinets, and circuit breakers to a common ground connection (discussed later). The ground wire does not carry any electrical current except during times of electrical fault.

Figure 14-8.Battery cables for use with NEC-approved systems must be made with approved building wire. The CEC does not have this restriction.

White, natural gray colored insulation

This cable wire may also be any color at all, other than green, provided the ends of the cable are wrapped with colored tape to clearly identify it as white. This wire carries current and is normally the negative conductor of the battery, PV, wind turbine, or inverter supply. The white wire is also bonded to the system ground connection, as detailed in Figure 14-8 and as discussed later in the text.

Red or other color

Convention requires that the red conductor of a two-wire system be the positive or ungrounded conductor of the electrical system. However, the ungrounded conductor may be any color except green or white.

The majority of DC systems are based on this standard color code. The AC side of the circuit uses a similar approach to color coding except that the ungrounded conductor(s) are generally black and red (for the second wire).

System Grounding

Grounding provides a method of safely dissipating electrical energy in a fault condition. Yes, that third pin you cut from your extension cord really does do something. It provides a path for electrical energy when the insulation system within an inverter or cable covering fails.

Imagine a teakettle for a moment. Two wires from the house supply enter the teakettle, plus a ground wire. During normal operation, the electricity flows from the house electrical panel via the ungrounded "hot" conductor to the kettle. Current flows through the heater element and back to the panel via the "neutral" conductor that is grounded. A separate ground wire connects the metal housing of the kettle (via the pesky third prong) to a large conductive stake driven into the earth just outside the house.

If the insulation or hot wire were to be damaged inside the kettle, it could touch the metal chassis. Because the chassis is bonded to ground (assuming you didn't cut the pin), electrical energy will travel from the chassis, through the ground wire to the conductive stake. This flow of current is unrestricted due to the bypassing of the element, causing overheating of the electrical wires. If it were not for the circuit breaker or fuse limiting this excessive current flow, a fire could start.

Electrical energy has an affinity for a grounded or "zero potential" object and will do whatever it takes to get to there. If there were no ground connection on the defective kettle's chassis, electricity would simply stay there until an opportunity arose to jump to ground. If you were to touch the kettle and simultaneously touch the sink or be standing on a wet surface, the electricity would find its path through your body. This is not a good situation.

In a similar manner, the entire exposed metal and chassis of the system components are bonded to a *common* ground point as illustrated in Figure 14-9. The white or negative wire of the system is also bonded to this point, saving us from adding a second set of fuses and a disconnection device to satisfy NEC/CEC requirements.

The size of the ground wire is determined by the NEC/CEC and is based in part on the size of the main over-current protection device rating, as shown in Table 14-2.

Lightning Protection

If you look carefully at Figure 14-6, you will notice two lightning arrestors attached to the main DC disconnection panel and the generator transfer panel. These devices contain an electronic gizmo known as a Metal Oxide Varistor (MOV), which connects between the DC +/- conductors and ground as well as the AC hot/neutral conductors and ground. A lightning arrestor is about the cheapest piece of insurance you can purchase to protect your power system. Connect one on every cable run that strings across your property. The system shown in Figure 14-6 has a roof-mounted PV panel and a generator located in a remote building. Cables running here and there can attract lightning on its way to ground potential. The arrestor "clamps" this voltage and passes it safely to the grounding conductor.

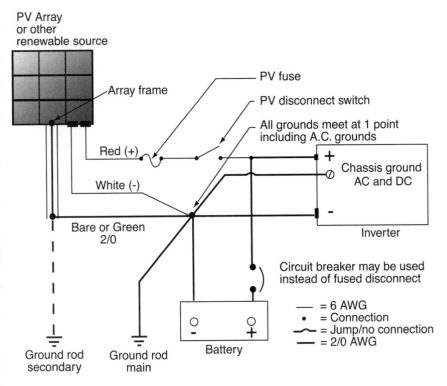

Figure 14-9. This view of a PV array, battery, and inverter DC interconnection shows the placement of over-current protection fuses and disconnect switch in the ungrounded (+) conductor. The negative (-) conductor (white) is bonded to ground and therefore requires no disconnection and over-current protection device. The ground wire (bare or green) connects at one point to a ground rod. A secondary ground rod may be used where the array or generator is a long distance from the main rod. Note the use of an interconnecting cable between both rods.

Size of Largest Over-Current Device	Minimum Size of Ground Conductor
Up to 60 A	#10 AWG
100 A	#8 AWG
200 A	#6 AWG
300 A	#4 AWG
400 A	#3 AWG

Table 14-2. This table shows the minimum size of the grounding conductor based on the largest over-current device supplying the DC side of the inverter.

Interconnecting the Parts

A mechanical layout plan of your desired installation will help you determine the material required and assist you in visualizing the layout of the power station. It is also important to determine what functions you require from the overall design. Some configurations include:

- grid-dependent system without battery backup;
- grid-interactive system with battery backup ;
- grid interactive-system with battery backup and generator emergency supply ;
- off-grid system with generator backup.

The number of configurations is extensive and it is not possible to cover every installation arrangement within these pages. Fortunately, the installation manuals for the inverter and equipment will assist with reconfiguring for custom-design requirements. By way of example, look at the schematic grid-interactive design using PV and wind power shown in Figure 14-10. For simplicity, no ground wiring is shown in this view. Refer to Figure 14-9 for grounding requirements.

PV Array

PV arrays may be mounted on a house roof, ground mount, or sun tracking unit. The interconnection of the modules is similar, with only one difference between the various locations. The NEC requires that every roof-mounted PV array be equipped with a device known as a *Ground Fault Interrupter* (GFI), which is shown in Figure 14-11. It automatically disconnects the PV array in the event of over-current, insulation, or water leakage fault that may cause overheating and a possible fire. A GFI is not required on ground, pole, or tracker mounts.

The first step in wiring your PV array is to determine the battery voltage you will be using. We discussed in Chapter 8, "Photovoltaic Electricity Production", how a standard module is manufactured for 12 or 24 V output. Chapter 8 also covered the steps required to increase PV array voltage by wiring modules in series as well as in parallel connection for increasing current flow. Figure 14-12 shows a group of four 12 V modules interconnected together in series to form a 48 V string. Each module is provided with a weatherproof junction box and knockout holes for liquid-tight strain-relief bushings. These bushings press into the hole in the junction box and are held in place by a retaining nut. The flexible cable is then passed between junction boxes and the series or parallel interconnection is completed. The schematic in Figure 14-10 shows two 24 V PV panels wired in series to create 48 Vdc.

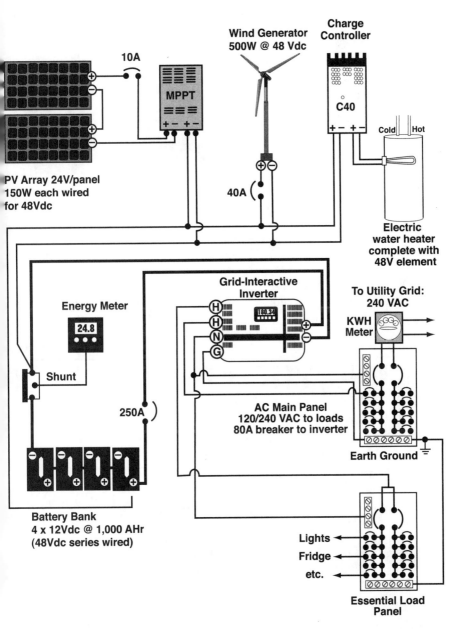

Figure 14-10. This is one of the most complex versions of grid-interactive design using PV and wind energy sources, battery backup, series battery charge control (with Maximum Power Point Tracking), diversion load charge regulation (for the wind turbine), and energy metering.

Be careful to check for proper wire gauge and type to be sure it is suited for outdoor, wet installation. Review the wire choice with your electrical inspector or the NEC/CEC code rulebook.

The output from a grouping of several modules then meets at a combiner box mounted on the rear of a sun tracker unit, as shown in Figure 14-13. The array comprises a total of sixteen 12 V modules which are connected in eight sets of two pairs. Each module is rated 75 W. The two modules are wired in series forming

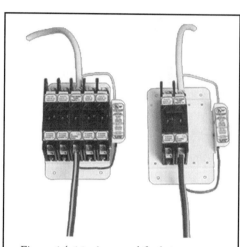

Figure 14-11. A ground fault interrupter provides protection against internal PV module faults that may cause overheating and fire. They are required when the PV array is roof mounted on a dwelling.

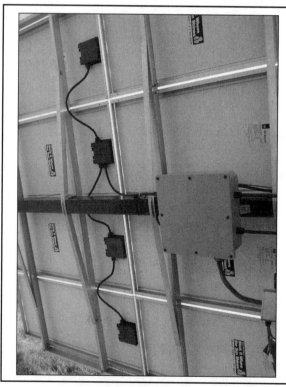

Figure 14-12. A PV array detail showing four 12 V modules interconnected in series, forming a 48 V string. The output of each of the four strings feeds into the combiner box shown at the center.

a 24 V set. The wires for the eight sets are then directed to the combiner box where all of the negative wires are connected together at one point. The positive leads are each directed to an individual fuse or circuit breaker (5 A rating per set). The output of the circuit breaker is then fed to the common positive (+) terminal of the series-wired voltage controller, as shown in Figure 14-10. The parallel connection can be completed using terminal strips to connect all of the appropriately sized wires to the main DC supply cable.

Digressing for a moment, let's look at how this works when fully connected:

a) *16 modules x 75 W per module = 1,200 W peak for the array*
b) *2 x 12 V modules wired in series = 24 V*
c) *Current from array = 1,200 W peak ÷ 24 V = 50 A maximum*

If we had simply left the array wired at 12 V and paralleled the 16 modules we would have had to deal with twice the current (100 A). Likewise, if the array were to be wired at 48 V, the current would be 25 A. Power in watts remains the same regardless of how we interconnect the modules. Higher voltage allows the use of a smaller conductor cable, reducing cost and energy losses.

Wind Turbine Connection
Connection of a wind turbine is essentially the same as for the PV array. The installation manual for each model will provide a connection point or suggest a wiring interconnection method. For example, Bergey Windpower recommends strapping a weatherproof junction box near the top of the tower, drilled to allow the feed/connection wires from the turbine to enter using a liquid-tight fitting to provide strain relief for the wires and keep the weather out.

A feed cable is then run down the tower to a second servicing junction box. The feed cable running the length of the tower (and finally to the house) should be installed in either a PVC plastic conduit or metal conduit secured to a tower leg with non-corrosive metal clamps. An equally acceptable alternative is a flexible cable/conduit combination known as teck cable. This material is more expensive than traditional conduit but installs in seconds and is just as strong. Use ultraviolet light-resistant cable ties to secure the teck lead wire to the tower leg.

Renewable Source to Battery Feed Cable
The electricity from the PV array (or wind turbine) must be routed to the battery room. Cables may be fed in conduit down the side of the house or wired overhead or underground if the generator is located some distance

Figure 14-13. A combiner and fuse box is fabricated using a weatherproof junction box and liquid-tight strain relief. The feed wires from the array enter the box and are parallel connected, with each positive lead passing through a series-connected fuse or circuit breaker.

away. Underground cable connections tend to be the most common because of simplicity and lower installation cost. The NEC/CEC has provisions for both direct burial cable and cable protected by conduit. In either case, a trench is dug, typically 18" (0.5 m) deep, between the source and the inverter/battery room or house. Where the cable exits the ground to enter the house or travel up the array/turbine support leg, it is necessary to use a length of conduit to protect the wire from damage. The cable is placed in the trench and covered with 6" (15 cm) of soil. At this point, an "underground wire" marking tape is placed into the trench. The intent of the tape is to warn anyone digging in the area that a buried cable lies below. The trench is then filled in.

Over-Current Protection and Disconnect Devices

When the energy source is connected to the inverter (grid-dependent) or batteries (grid-interactive), the NEC/CEC requires that an over-current protection and disconnection device be installed. This may be done in one of two ways: you can purchase individual fused disconnect switches for each source, as shown in Figure 14-14; or you can have auxiliary circuit breakers added to the main battery/inverter disconnect box. The latter was chosen for the system outlined in Figure 14-1.

Every manufacturer of inverter-based systems provides some form of integrated panel as shown in Figure 14-1. Although initial cost will be higher for an integrated wiring panel, final cost will be the same or lower compared to purchasing the parts "a la carte" and wiring them together. In addition, any problems with incompatibility or "finger pointing" are eliminated when the complete system is provided by one manufacturer.

Battery Wiring

As you learned in Chapter 10, "Battery Selection and Design", wiring batteries in series increases the voltage (Figure 14-15A), and connecting them in parallel increases the capacity (Figure 14-15B).

Figure 14-14. Individual fused disconnection boxes, such as the one shown here, may be used for each renewable source of energy or may be combined within the main disconnect chassis.

Battery manufacturers offer many voltages and capacity ratings to suit your specific requirements. The most important considerations for installation are voltage, capacity in amp-hours (Ah), and distance from the inverter. You will have to determine your essential load power requirements and the length of time you expect these items to operate during a blackout before selecting a battery configuration. Refer to Appendix 7, "Electrical Energy Consumption Worksheet", to calculate the backup energy requirement and resulting battery capacity.

As with integrated wiring panels, some manufacturers are offering inverter/ battery combination packages to take the guesswork out of these calculations. Beacon Power offers its grid-interactive *Smart Power M5* and Outback Power Systems provides the *PS1-GVFX3648* system for outdoor installation.

As the size of the battery bank physically increases (providing power for long-duration blackouts) it becomes more difficult to use short wire lengths to feed the inverter. Plan the battery room layout using graph paper and cutouts of the selected battery model to determine the best physical layout, paying close attention to the cable connections and the length of the runs. Be sure to allow for battery disconnection and over-current protection within the wiring layout. Refer to Chapter 10 for further details regarding series/parallel battery theory.

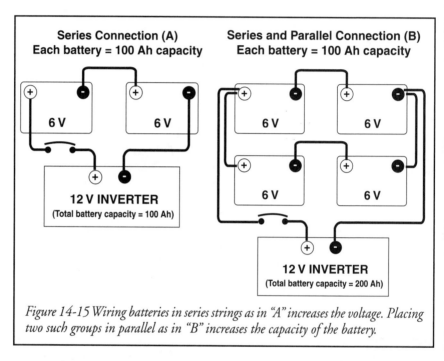

Figure 14-15 Wiring batteries in series strings as in "A" increases the voltage. Placing two such groups in parallel as in "B" increases the capacity of the battery.

Battery capacity and physical size can become so large that the length from one end of the battery bank to the inverter may exceed the manufacturer's recommended cable length for inverter operation. To circumvent this problem, interconnection wires should be bumped up to the largest size possible (#4/0) with the positive and negative leads taped together along their lengths to reduce cable inductance. Sometimes there is only so much you can do!

Battery Voltage Regulation

Battery voltage is regulated using a charge controller, which was discussed in Chapter 11, "DC Voltage Regulation". The charge controller may be connected as either a series regulator or a diversion regulator. The series regulator is the simplest arrangement, simply turning the PV input to the batteries on and off based on their state of charge. If your system includes a wind turbine, it's almost always necessary to maintain a load on the generator, requiring a shunt or diversion load arrangement.

Figure 14-10 illustrates a system designed with both a series controller equipped with a maximum power point tracking feature and a diversion charge controller with an electric water heater operating as the diversion load.

Series Regulator

Series regulation is by far the simplest and least expensive system to install. The downside with series regulation is that excess energy is wasted when the unit is regulating battery voltage. Morningstar Corporation indicates that series regulation adds less than 5% to the cost of a mid- to large-sized PV array and battery system. Maximum power point tracking options will increase this expense somewhat.

In Figure 14-10, a PV array has been configured by series wiring two 24 V panels to provide a nominal output of 48 volts DC. As you learned in Chapter 8, the open circuit voltage (PV panel in full sun without a connected load) will develop approximately 34 volts for a nominal 24 volt model. If this array were connected directly to the 48 volt battery bank, the output voltage of the PV array would match that of the battery.

The PV array will generate maximum power at only one point on its current/voltage rating. It is certain that the battery voltage will not coincide with the maximum power point of the PV array. The series controller connected to the PV array contains a maximum power point tracking feature which allows the array voltage to track the maximum power point. The output of the controller "converts" this input power to a level that maximizes battery charging capability. In addition, the controller will automatically taper the charging current as illustrated in Figure 11 -7.

Figure 14-16 shows the connection details for a Xantrex Technology C-40 model charge controller which is not equipped with maximum power point tracking. The connection details can't get much simpler than this. Note that circuit breakers (or fused disconnect switches) must be used in each leg of an ungrounded wire (positive lead) fed by a source of voltage.

Shunt or Diversion Regulator

Figure 14-10 shows a Xantrex model C40 charge controller configured for diversion load control. This configuration is required when grid failure occurs and the battery bank is fully charged. When these conditions are ac-

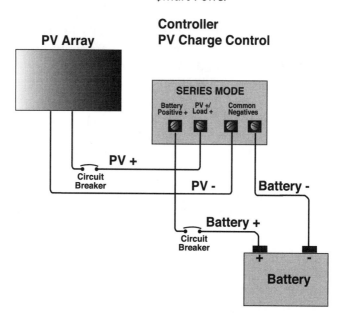

Figure 14-16. A charge controller may be connected for series or diversion regulation. The series regulator connection shown above is the simplest and least expensive arrangement.

companied by strong wind, the wind generator will produce more energy than is required by the system. As discussed previously, wind and micro hydro turbines often require an electrical load to be connected at all times.

When the batteries are full their voltage will rise, telling the PV array series controller to stop charging (Figure 11-7). This releases the electrical load on the PV array. If the electrical load were released from the wind generator, the blades could possibly enter an over speed mode, destroying the unit. To prevent this from happening, the diversion charge controller "shunts" excess energy from the turbine to the diversion load (in this case a water heater), maintaining the required load.

Review Chapter 11 for details on plumbing and electric heating element swapping. The standard 120/240 V elements of an electric water heater used as a diversion load **MUST** be replaced with new ones rated at the nominal wind turbine/battery bank voltage. Ensure that internal thermostats and over-temperature protection devices within the water heater are bypassed, either by removing them from the wiring circuit or installing a suitable heavy-gauge jumper wire. You do not want an internal thermostat opening, thereby disconnecting the load from the turbine.

Air-heating elements such as the model shown in Figure 14-18 can be

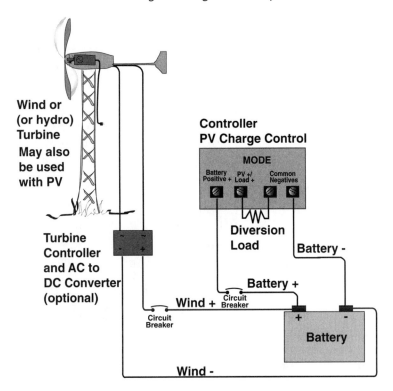

Figure 14-17. With a shunt or diversion regulator, the charge controller is connected to a diversion load such as a hot water heater or electric air-heating element.

used as a diversion load. However, air heating tends to be wasteful, as the majority of the heat may be generated outside the heating season. If an air-heating load is used and the byproduct heat is not required, consider mounting the heating element in a rainproof outdoor chassis.

The electrical rating of the diversion load should be 25% greater than the capacity of the devices feeding it. For example, the wind turbine shown in Figure 14-10 is rated 500 W and the PV array does not contribute to the diversion energy because it is equipped with its own series charge controller. Therefore, the diversion load rating should be 500 W plus a 25% safety factor, requiring an electric heating element of 625 W capacity. The rating of the diversion controller must also be calculated to ensure that it is able to supply the maximum power to the diversion load. The charge controller rating is calculated by dividing the diversion heater rating (in watts) by the nominal system voltage:

625 W diversion load ÷ 48 V system voltage = 13 amp diversion controller rating

This is well within the Xantrex C40 load rating capacity of 40 amps.

Directing the excess heat to a hot tub or spa is another good place to "dump" excess energy. Use caution when connecting the diversion elements in this application. Both Underwriters Laboratories and Canadian Standards Association safety standards require the use of safety current collectors and GFI protection in spa systems[1].

Figure 14-18. *Absorbing large amounts of diverted electrical energy is easy if heat enters the equation. An air heating element such as the model shown here will absorb up to 1 kW of excess energy and provide some home heating to boot.*

The Inverter

DC Input Connection

The inverter is the heart and brains of the renewable energy system. Inverters are also pretty darned heavy, and if your system is designed to operate 240 V house loads it may be necessary to have two inverters "stacked" together to generate this voltage. Figure 14-20 shows a prewired power panel from Xantrex Technology Inc. that contains:

• two 4,000 W sine wave inverters
• two series-wired charge controllers
• dual 250 A over-current disconnect units
• AC wiring chassis for generator and house panel connection

---

[1]   UL Standard UL 1795 and CSA standard C22.2 #218.1 require the use of current collectors to prevent shock hazard in the event of heater failure. 120 V spa-heating elements may be purchased from your local pool and spa supply store. Consult with the applicable standards before connecting the diversion controller in this situation.

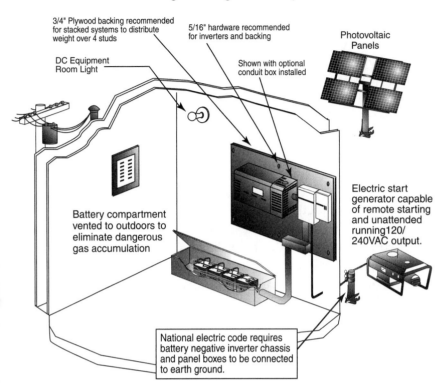

3/4" Plywood backing recommended for stacked systems to distribute weight over 4 studs

5/16" hardware recommended for inverters and backing

Photovoltaic Panels

DC Equipment Room Light

Shown with optional conduit box installed

Battery compartment vented to outdoors to eliminate dangerous gas accumulation

Electric start generator capable of remote starting and unattended running120/ 240VAC output.

National electric code requires battery negative inverter chassis and panel boxes to be connected to earth ground.

Figure 14-19. This drawing details many of the issues related to the DC connection of the batteries, inverter, charge controller, and disconnect and over-current protection system. Consult with the NEC/CEC and your electrical inspector to determine the details specific to your installation. (Drawing based on Xantrex Technology Inc. installation designs)

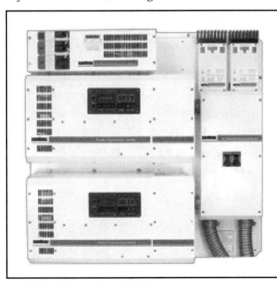

Figure 14-20. This prewired electrical system is compact, neat, properly installed and, most importantly, certified, which reduces problems with sometimes wary electrical inspectors. (Courtesy Xantrex Technology Inc.)

Purchasing your system prewired like the ones shown in Figures 14-1 and 14-20 is a wise decision. As the panels are assembled at the factory and approved to applicable safety standards, there is no problem satisfying your electrical inspector. In addition, costly errors in wiring are eliminated.

If you wish to complete your own wiring a la carte style, your electrical inspector will require a simplified wiring schematic such as the one shown in Figure 14-10. Note that this schematic drawing is in no way a full interpretation of the NEC/CEC code rules, but it does provide you with a basis for discussion with your inspector.

AC Output Connection

So far we have categorized renewable energy systems as being either on or off the electrical grid. While this generalization is accurate, what it does not reflect are the dozens of ways the systems can be wired and interconnected together. Some of the more common system configurations are:

- on-grid, no battery storage
- on-grid, battery storage for short-term outage protection
- on-grid, battery storage and genset for long-term outage protection
- off-grid, no genset backup power source
- off-grid, genset power source

(Note that each system may be connected either for 120 V AC operation only or for 120/240 V AC operation, typically using two inverters.)

This basic combination of systems produces ten different wiring methods with many more designs being possible. It will be necessary to decide which system configuration suits your requirements and then consult with the various equipment manuals to determine proper wiring configurations. For the unit shown in Figure 14-10, we will assume the gird-interactive mode configuration typical of many installations.

The AC side of the inverter will be well known to any electrician. Single-phase voltage at 120 volts is supplied by one black "hot" wire and a white "neutral" return wire. A safety ground is also required as discussed above. If the configuration includes a 240 volt supply, a second black "hot" wire will be provided. The first and second hot wires are commonly referred to as line one and line two respectively.

During periods of renewable energy production, electricity is fed from the hot (Line 1) terminal of the inverter to the house AC main panel, which supplies household loads with any surplus being sold to the electrical utility. The importing and exporting of electricity to and from the utility is determined by the direction of power flow recorded by the revenue meter. The system will always provide power to the household loads first, with the surplus being sold to the utility second.

Figure14-10 shows the grid-interactive inverter wired for 120 volt connection to the AC mains panel. A second inverter may be connected or "stacked" to double the capacity of the renewable energy system and work in tandem with the first inverter, feeding 240 volt power to the AC mains panel.

Regardless of which system you wish to install, your electrical inspector will want to see a wiring diagram detailing the various connections, particularly when the electrical grid is involved. For an example, see the simplified schematic shown in Figure 14-10. You will have to include details such as wire sizes and type, fuse and circuit breaker ratings, and any other details requested by the inspection authority.

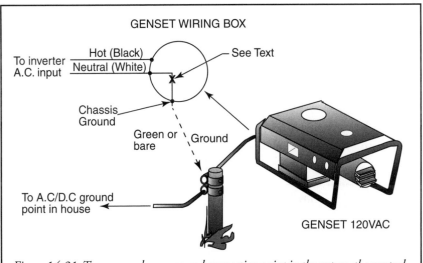

Figure 14-21. To ensure only one ground connection point in the system, the neutral-to-ground bond inside the generator will have to be removed as shown here.

The Generator

Generators supplied in North America are available with 120/240 V split phase or 120 V only output configurations. If your renewable energy system is wired to provide 120 V only, having your supplier prewire the generator for 120 V will make installation simpler. The same is true for 240 V systems.

The requirement of having all ground and neutral points bonded at one location (see Figure 14-21) will necessitate the removal of the ground-to-neutral connection inside the generator wiring chassis if the unit is part of a grid-interconnected system.

Energy Meters

Earlier we discussed how an electrical energy meter records the number of electrons flowing into or out of the battery (current exported or imported). A device known as a shunt (detailed in Figures 14-10 and 14-22) converts this current flow into a signal that can be measured by the meter, allowing the calculation of energy consumed and generated. The shunt is usually mounted inside the breaker chassis box, ready for wiring in series with the battery bank negative terminal.

If you are purchasing a prewired system it is well worth a few extra dollars to have a shunt and energy meter prewired. An example of the energy meter is shown in the top right corner of Figure 14-1. These handy little devices record energy consumed, produced, and, most importantly, battery state of charge, which will help prevent damaging over-discharge conditions.

If you are considering wiring your own system refer to Figure 14-22, which details the connection of the shunt. Refer to the specific energy-meter wiring-connection diagrams for the balance of the system connection. Note that where the small-gauge signal wire (typically #22 to #28 AWG) cable exits a hole in the breaker chassis, an anti-rubbing bushing must be added to prevent damage.

As with prewired panels it is wise to have the shunt installed by the manufacturer, as the large wire gauge and tight chassis conditions make installation after the fact almost impossible.

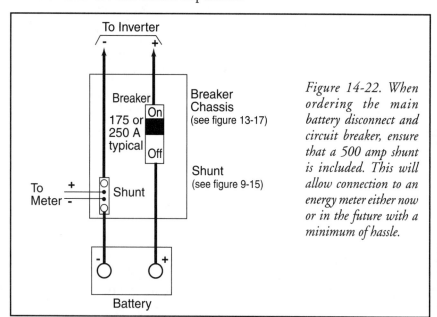

Figure 14-22. When ordering the main battery disconnect and circuit breaker, ensure that a 500 amp shunt is included. This will allow connection to an energy meter either now or in the future with a minimum of hassle.

15
Living with
Renewable Energy

Solar Thermal Systems

Once a solar thermal system has been installed and commissioned, very little maintenance as required. For the majority of homeowners, sweeping snow from the collectors (assuming they are safely accessible) is the limit of day-to-day maintenance.

The majority of systems installed throughout North America use non-toxic antifreeze known as propylene glycol to transfer heat from the collector panels to the storage tank in the house. Propylene glycol is rated for a maximum operating temperature of 325°F (163°C), which is well above the normal operating temperature of the system. During the summer months, particularly when people are on vacation, stagnation of the propylene glycol may occur due to the low heating load of the home. Should stagnation continue for extended periods of time, the glycol solution will become "sticky," and gum up the works.

It is recommended that a trained service technician analyze the glycol solution every second year. During this test, all system components should also be checked to ensure peak operating efficiency.

Solar Electric Systems

Well, let's get ready to break out the champagne and "throw the switch." All the wiring is done, circuits are checked and rechecked, and the electrical inspector has given the OK. It's time to get things running. So where do you start? It can't really be that simple. After all, there's a lot of equipment involved.

Getting Started

The best place to start is with safety and also with some background on operating a system. Each source of renewable energy, whether it's a wood stove or PV panels, has an operating guide supplied with it. All manufacturers provide a list of the safety tools and supplies you will need when working with their products. This equipment has been covered in the previous chapters, so go back and review the material as required. The most obvious requirements are common sense and the ability to learn as you work with the various components.

Once all of the systems are checked out in accordance with the manufacturers' requirements, we can start things moving. Obviously, the first thing to do is to energize the system as discussed in the following paragraphs:

Grid-Dependent (non-battery based) Systems

The steps required to operate a PV-based grid-dependent system are ridiculously simple. The main disconnection circuit breakers are engaged (turned on) for the AC and DC/PV side of the system in accordance with the manufacturer's instructions. Once the sun starts to shine, you should be up and running.

Figure 15-1. The coal-fired grid may become a thing of the past if sufficient renewable energy and efficiency measures are adopted by society at large. (Courtesy Ontario Clean Air Alliance)

If the installation includes an energy meter, you should see power flowing into the grid (sell or export mode) depending on the size of the PV array and the amount of sunlight present. At night the system should not generate any power at all. Heavy cloud, overcast skies, fog, and shading from trees or other obstructions will reduce power output dramatically. If the PV array is on a fixed mount, early morning or late afternoon sunshine will strike the panels on a glancing angle, also reducing energy output.

Maximum power output will be coincident with "solar noon," when the sun is perpendicular (or at right angles) to the array surface. High levels of humidity or dust in the air will reduce energy production. With practice you will learn to recognize the variations in solar illumination that cause energy output to fluctuate. Aside from cleaning off dust, dirt, and occasional bird droppings, the PV panels will require virtually no maintenance.

If the PV array is mounted on a frame that allows seasonal adjustment, a good rule of thumb is to set the winter angle at your latitude plus 15° to the vertical and the summer angle at latitude -15°.

Grid-Interactive (battery backup) Systems

Grid-interactive systems have all of the same operational installation concerns as the grid-dependent systems discussed above. When the utility power fails, the grid-interactive design will immediately operate your home in off-grid mode. If there is any area that a beginner will have problems with it is relating the amount of energy consumed to the amount of energy stored in the battery bank. As discussed in Chapter 10, over-discharging your battery bank is the quickest way to destroy it, giving you inadequate performance along the way.

Step 1. Record the specific gravity for each battery cell prior to initial operation unless you are using maintenance-free or NiCad batteries, in which case this step is not required. Recording specific gravity will provide a basis for determining the state of the batteries upon receipt and may be required if you have any warranty issues. Use a felt pen or marking plate to give each individual cell a number. If your system is 12 V, you will have 6 cells. Likewise 24 V and 48 V systems will have 12 and 24 cells respectively per battery bank. If you have multiple sets of batteries wired in parallel such as those shown in Figure 10-19, you will have double the number of cells in your bank.

Step 2. Energize the main circuit breaker connecting the batteries to the inverter. At this point, the inverter will be ready for startup. Follow the

manufacturer's suggested startup procedure to get the unit online. You will need to set up the following basic parameters:
- battery charger rate;
- battery bank capacity;
- battery type;
- over-discharge protection;
- search mode sensitivity.

More advanced inverters require you to set many more options, and this can be done initially or on an as-needed basis. For first-time operators, this would be a good time to have your system dealer available to do a bit of "hand holding."

Step 3. If your inverter is equipped with a "search mode" option, set the search sensitivity. This setting adjusts the current threshold required to bring the inverter out of the low-power "sleep" mode and into full power operation. As we learned in Chapter 1.3, the search mode works by sending a brief pulse of power into the power lines. When all circuits are turned off, the flow of electricity is stopped. At the instant the circuit is closed (when a light switch is turned on, for example), the current pulse travels through the load and back to the inverter, signalling it to activate. Sleep mode saves considerable electrical energy when the inverter is not required, for example when everyone is sleeping.

Search mode can be tricky to adjust as the sensitivity differs depending upon which load is activated. For example, an 8 W CF lamp uses far less power than a toaster. Following the inverter manual instructions, perform this adjustment using a variety of electrical loads in the house.

Step 4. Program the energy meter according to the manufacturer's instructions. Typical data required will include the battery bank voltage and capacity (in amp-hours). With all renewable source supplies turned off, the inverter will be drawing power of varying amounts depending on the electrical loads activated at that time. Try turning a light on or off to see the amount of current draw. Each load will require more or less current (in amps), causing the meter to register accordingly.

Step 5. Close the circuit breaker for the first renewable source supply. The input power will vary depending on whether it is sunny and/or windy outside. The energy meter should now start to record the energy going into the battery. A common problem at startup is the current

Figure 15-2. Saving energy is always more economical than generating additional energy. This is true whether you are talking about an urban homeowner or the electrical utility.

shunt wiring (Figure 14-22) that is connected to the meter backwards. A quick test is to turn off all electrical loads and monitor the meter when energy is being produced. The meter should indicate a "battery charging" condition with the capacity moving slowly towards the full state. Repeat this step for each renewable source.

Step 6. Close the circuit breaker for the generator feed inputs. Start the generator (if one is installed) either manually or by activating the autostart function in accordance with the inverter instruction manual. Once the generator has started and stabilized for about twenty seconds, the inverter will "click over" to battery-charge mode.

When this occurs, the batteries should start to charge at the inverter's maximum rate (or as programmed). The energy meter will record the resulting power input and charging will progress.

While generator charging is in progress, verify that the house electrical power is active. All lights and appliances should be able to run as if they were running on the inverter. Note that the generator power may add a "flicker" condition to lights. This is completely normal operation.

Step 7. Take time to read the energy meter manual to understand the relationship between the meter and batteries. Remember that the meter is only a "guesstimate" of actual battery state of charge.

Up and Running

The real beauty of a renewable energy system is its ease of operation once everything is commissioned and running. Correct operation and proper attention to the system will prolong its life. The most neglected part of a system is the batteries, as those unfamiliar with renewable energy systems often do not realize how much energy they are using or understand the relationship between personal energy use and battery depth of discharge.

If there is any point that needs to be made clear from the beginning it is that you do not have a line of credit with your battery bank. There is only so much "cash" in the savings account. Use it up and that's it; the lights go out. Keep draining the bank continually and the bank will close permanently.

Read the battery manual to ensure that you understand how your energy usage, energy meter, and battery specific gravity work together. CHECK THE SPECIFIC GRAVITY EVERY DAY, UNTIL YOU UNDERSTAND THIS RELATIONSHIP. Once you fully understand the energy meter/specific gravity interaction, you need to check the battery bank only once a month. But until then, use care and follow these steps to the letter:

- Battery specific gravity will read abnormally low immediately after a full day of charging. This is because of tiny gas bubbles suspended in the electrolyte, causing the density of the fluid to decrease. (Gas bubbles are lighter than water or sulfuric acid.) Wait at least one or two hours after charging before taking the specific gravity reading.
- Likewise, specific gravity reading should not be taken when the batteries are under a full load. Wait until the fridge stops or the microwave is not being used before taking readings.
- Battery electrolyte specific gravity readings that are outside room temperature range will need to be "corrected" in accordance with the manufacturer's data.
- Use a quality hydrometer to take the readings. Look at the scale from straight on to ensure that the reading is correct.
- Compare the specific gravity readings with the battery depth-of-discharge chart in Chapter 10, Table 10-1. If the battery bank indicates that it is 90% full (i.e. 10% depth of discharge) and the energy meter agrees, great! If not, the meter may need some fine-tuning if the readings are too far out of whack. You may wish to look at a figure called the *charge efficiency factor* setting. This is a fancy term for the fudge factor to help calibrate the meter-to-battery setting. Decrease the efficiency factor if the meter thinks the batteries are charged more fully than they really are.

Figure 15-3. Saving energy is not limited to your home. Hybrid vehicles such as this model from Toyota will reduce your automotive energy bill by 50% and smog-forming emissions by 90%.

Likewise, increase the efficiency factor if the meter reads lower than the specific gravity reading.

- As time goes by, the readings of even the best energy meter will fall out of step with the batteries. At this point, read the meter manual to determine how to reset the meter to agree with the battery capacity. This step is normally accomplished when the batteries are fully charged immediately after an equalization charge, approximately four times per year.
- Speaking of equalization, watch the specific gravity between cells. The readings should all be approximately the same. If the cells are starting to get out of balance by a reading of more than 0.010 (ten points on the scale), it's time to equalize.
- Equalization may be accomplished using grid power, a generator, or better yet, the PV panel or the wind turbine on sunny/windy days. Whatever method you choose, it is necessary to "program" the charge controller and/or inverter to start the equalization process. Many charge controllers have a single button which increases the battery maximum charge voltage, thus enabling equalization mode. The inverter will often have a

similar button or setting which is activated when equalization is started from a generator.

- If you use Hydrocaps with your battery bank **YOU MUST** remove them during equalization.

What Else?

There really isn't much else to think about, especially for grid-dependent systems. Equipment manufacturers will give you a list of yearly maintenance steps to follow. Some are absurdly simple, such as spring-cleaning the PV cells to wash off any dirt on the glass. Wind turbines require an inspection, although even this can be pretty simple. Bergey WindPower suggests checking to see if the unit is turning once every year….Honestly, it's right there in the manual!

Of course there are some tricks you will learn over time that help make living with renewable energy easier:

- Watch the battery specific gravity or energy meter when you have a long-term blackout of one day or longer. This is where many newcomers to renewable energy get into trouble with their batteries. Recognize when the batteries are depleted enough to warrant shutting the system down or running a generator.

- During severe lightning storms, consider furling your wind turbine to limit mechanical stress during the storm. Consult the manufacturer's manual for additional details.

- Make seasonal adjustments to a manual PV array: latitude +15E towards the vertical for winter; latitude -15E for summer.

- Remember never to smash ice and snow off the PV panels. Simply brush off the top layer of snow with a squeegee or brush. The sun will quickly take care of the rest.

- Rest easy knowing that you are not the only one living lightly on the planet. Support abounds from dealers, like-minded neighbors, the web, and publications such as *Private Power Magazine* and *Home Power* (www.homepower.com). Renewable energy is here to stay.

Probably the most important issue of all is to enjoy your handiwork and marvel at the elegant simplicity of it all.

Figure 15-4. Heating your pool or hot tub with solar energy eliminates fossil-fuel heating, extending the swimming season without increasing the operating cost.

16
Conclusion

Energy *is* the life breath of society. North Americans have been blessed with seemingly unlimited natural resources and a relatively small population density in comparison to the rest of the world. Multiple generations of double-digit economic growth have created the most luxurious, wealthy, and wasteful society in all of humanity. As we enter the 21st century most people still believe that growth-based capitalism is the only way forward.

Continued growth requires more energy and natural resources as inputs, creating more products and pollution as outputs. And what does all of this increased stuff give us? Is your life better off with more and larger cars, a bigger home, and higher income? Stop and think about this for a minute. A lifestyle developed to keep pace with society's idea of what you should have increases living expenses, requires two incomes, and creates higher levels of stress and little if any downtime. Certainly a guilt-ridden, hastily planned "holiday" to take the kids to Disneyland is not going to do anything for your soul.

The need to consume more and more stuff to feed our empty hearts and expanding egos cannot continue unabated forever. As we voraciously consume our natural capital and pollute the very ecosystem that sustains life, we are behaving like a disease that consumes its host. Might humans be the cancer of this planet? Without a doubt we are. This is not doom-saying or apocalyptic prophesy; it is the truth. The fact is that at the some point in the future, sooner or later, society's endless consumption and destruction of our ecosystems will upset the delicate balance we now choose to neglect, with the result that human life will be relegated to history.

Will sustainable development work and get us out of this mess? If all of society adopted the strategies in this book would we stop the slide into the abyss? I don't know. What I do know is that human beings are today's caretakers of this Earth and we have a responsibility to pass it on to future generations as carefully preserved as possible.

We rely on our technology to solve all of the problems that modern society creates. If we don't have an answer now we postpone decisions until sometime in the future. The fuel bundles used in a nuclear reactor generate power for a period of weeks, yet the waste remains hazardous for over *10,000 years*, more than twice as long as the existence of civilization. Society's collective answer to this dilemma is to try and bury the stuff and hope that future generations can figure out what to do.

The practices and by-products of modern society—smog, greenhouse gases, mercury poisoning, insecticides, trash, inefficient buildings, nuclear weapons, mass production farming, the destruction and waste of water, biological warfare, resource depletion—are stacking the deck against us.

Sustainable development and clean energy technologies may only slow our demise, but as the custodians of this wondrous planet, it is our responsibility to ensure that the clock does indeed keep on ticking.

To laugh often and love much; to win the respect of intelligent persons and the affection of children; to earn the approbation of honest critics and to endure the betrayal of false friends; to appreciate beauty; to find the best in others; to give of one's self; to leave the world a bit better, whether by a healthy child, a garden patch, or a redeemed social condition; to have played and laughed with enthusiasm and sing with exultation; to know that even one life has breathed easier because you have lived — this is to have succeeded."

Ralph Waldo Emerson, (1803 -1882)

Appendix 1
Cross Reference Chart of Various Fuels Energy Ratings

Fossil Fuels and Electricity

Heating Fuel	BTU per Unit
Heating Oil	142,000 BTU/gallon (38,700 kj/L)
Natural Gas	46,660 BTU/cubic-yard (37,700 kj/m3)
Propane	91,500 BTU/gallon (26,900 kj/L)
Electricity (resistance heating)	3413 BTU/kWh (3600 kj/kWh)
Coal (air dry average)	12,000 BTU/LB (27,900 kJ/kg)

Renewable Energy Heating Fuels

Heating Fuel	BTU per Unit
Shelled Corn	7000 BTU/lb (16,200 kJ/kg)
	14,000,000 BTU/ton (12,700 MJ/tonne)
Firewood by weight (all types) (note 1)	8000 BTU/lb (18,500 kJ/kg)
Hardwood Firewood by volume: (note1)	
Ash	25,800,000 BTU/cord (27,200 MJ/cord)
Beech	28,900,000 BTU/cord (30,500 MJ/cord)
Red Maple	22,300,000 BTU/cord (23,500 MJ/cord)
Red Oak	27,200,000 BTU/cord (28,700 MJ/cord)
Hybrid Poplar	18,500,000 BTU/cord (19,500 MJ/cord)
Mixed Hardwood (average)	27,000,000 BTU/cord (30,000 MJ/cord)
Mixed Softwoods (average)	17,500,000 BTU/cord (18,700 MJ/cord)
Wood Pellets	20,700,000 BTU/ton (19,800 MJ/tonne)
Biodiesel	128,000 BTU/gallon (35,500 kJ/L)

Note 1:
All firewood has the same heating or carbon content per pound or kilogram of mass. However, the density of softwoods are much lower owing to increased air and moisture content, resulting in lower BTU content per unit mass.

Appendix 2
Typical Power and Electrical Ratings of Appliances and Tools

Appliance Type	Power Rating (Watts)	Energy Usage per Hour, Day or Cycle
Large Appliances:		
Gas clothes dryer	600	500 Wh per dry cycle
Electric clothes dryer	6,000	5 kWh per dry cycle
Hi efficiency clothes washer	300	250 Wh per wash
10 yr. old vertical axis clothes washer	1,200	720 Wh per wash
10 yr. old refrigerator	720	5 kWh per day
New energy efficient refrigerator	150	1.2 kWh per day
10 yr. old chest deep freeze	400	3 kWh per day
New energy efficient chest deep freeze	140	0.9 kWh per day
Dish washer "normal cycle"	1,500	800 Wh per cycle
Dish washer "eco-dry cycle"	600	300 Wh per cycle
Portable vacuum cleaner	600	600 Wh per hour
Central vacuum cleaner	1,400	1.4 kWh per hour
Air conditioner 12,000 BTU (window)	1,200	1.2 kWh per hour
AC submersible well pump (1/2 hp)	1,150	200 Wh per cycle
DC submersible well pump	80	160 Wh per day
DC slow pump (includes booster pump)	80	160 Wh per day

Appliance Type	Power Rating (Watts)	Energy Usage per Hour, Day or Cycle
Small Appliances:		
Microwave oven (0.5 cubic foot)	900	0.9 kWh per hour
Microwave oven (1.5 cubic foot)	1,500	1.5 kWh per hour
Drip style coffee maker (brew cycle)	1,200	1.2 kWh per hour
Drip style coffee maker (warming cycle)	300	0.3 kWh per hour
Espresso/cappuccino maker	1,200	300 Wh per cycle
Food processor	300	50 Wh per cycle
Coffee grinder	100	10 Wh per cycle
Toaster	1,200	150 Wh per cycle
Blender	300	50 Wh per cycle
Hand mixer	100	10 Wh per cycle
Hair dryer	1,500	200 Wh per cycle
Curling iron	600	100 Wh per cycle
Electric tooth brush	2	50 Wh per day
Electric iron	1,000	1 kWh per hour

Appendix 2 Continued

Appliance Type	Power Rating (Watts)	Energy Usage per Hour, Day or Cycle
Electronics:		
Television –12 inch B&W	20	20 Wh per hour
Television –32 inch color	140	140 Wh per hour
Television –50 inch hi definition	160	160 Wh per hour
Satellite dish and receiver	25	25 Wh per hour
Stereo system	50	50 Wh per hour
Home theater system (watching movie)	400	1 kWh per movie
Cordless phone	3	72 Wh per day
Cell phone in charger base	3	72 Wh per day
VCR/DVD/CD component	25	25 Wh per hour
Clockradio (not including inverter waste)	5	120 Wh per day
Computer "Tower"	60	60 Wh per hour
Laptop computer	20	20 Wh per hour
15 inch monitor	100	100 Wh per hour
15 inch flat screen monitor	30	30 Wh per hour
Laser printer (standby mode average)	50	50 Wh per hour standby
Laser printer (print mode)	600	600 Wh per hour printing
Inkjet printer (all modes)	30	30 Wh per hour
Fax machine	5	120 Wh per day
PDA charging	3	72 Wh per day
Florescent desk lamp	10	10 Wh per hour

Appendix 3
Resource Guide

Energy Efficiency Councils and Societies:

American Council for an Energy Efficient Economy
Website: www.aceee.org
Phone: 202-429-8873
Publishes guides' comparing the energy efficiency of appliances

American Solar Energy Society
Website: www.ases.org
Phone: 303-443-3130
ASES is the United States chapter of the world Solar Energy Society. They promote the advancement of solar energy technologies.

The American Wind Energy Association
Website: www.awea.org
Phone: 202-383-2500
The AWEA is the trade association for developers and manufacturers of wind turbine and associated equipment and infrastructure.

California Energy Commission
Website: www.energy.ca.gov
Phone: 916-654-4058
The CEC is the strongest supporter of grid inter-connected renewable energy systems in North America. Their website explores what is happening in California in this regard.

Canadian Standards International
Website: www.csa-international.org
Phone: 416-747-4000
Develops standards for the Canadian marketplace. Tests and administers safety certification work in North America.

Canadian Wind Energy Association
Website: www.canwea.ca
Phone: 800-992-6932
The CAWEA vision is to have 10,000 MW of wind power systems installed in Canada by 2010. They promote all aspects of wind energy and related systems to the industry.

David Suzuki Foundation

Website: www.davidsuzuki.org
Phone: 614-732-4228
Dr. Suzuki is a lecturer and TV broadcaster promoting energy efficiency, global climate change and ocean sustainability. The website provides links and publications on all manner of environmental sciences.

Electro Federation of Canada

Website: www.micropower-connect.org
Phone: 905-602-8877
The Electro Federation is a consortium of electrical manufacturers working in many disciplines of electrical engineering and sales. The micropower-connect division is dealing with small (<50kW) distributed energy producers interconnecting to the grid in Canada.

Energy Star

Website: www.energystar.gov
Phone: 888-782-7937
Their website reviews energy efficient computers and electronics.

The Green Power Network

Website: www.eren.doe.gov/greenpower
This website describes the status of utility interconnection guidelines on a state-by-state basis. Also provides information on where to purchase green electricity when connected to the grid.

National Renewable Energies Laboratory (NREL)

Website: www.nrel.gov
Phone: 303-275-3000
The NREL is the national renewable energy research laboratory in the United States.

Natural Resources Canada

Website: www.nrcan.gc.ca
Phone: N/A
The Government of Canada hosts this website which includes the office of energy efficiency. Many resources are presented in this fact filled site.

Rocky Mountain Institute

Website: www.rmi.org
Phone: 970-927-3851
The Rocky Mountain Institute is a think tank regarding all energy efficiency issues. Their website contains a great deal of source information for books and applied research.

Underwriters Laboratories Inc.
Website: www.ul.com
Phone: 847-272-8800
Develops standards for the United States marketplace. Tests and administers safety certification work in North America.

Trade Publications and Magazines:

Home Power Magazine
Website: www.homepower.com
Phone: 800-707-6585
This magazine bills itself as "The hands-on journal of home-made power". Based in Oregon, the magazine deals primarily with a south and west coast flavor. Extensive details related to producing alternate energy.

Private Power Magazine
Website: www.privatepower.ca
Phone: 800-668-7788
Private Power is a new comer to the Canadian market place. Focuses on hybrid installations required in the North.

Alternative Power Magazine
Website: www.altpowermag.com
Phone: N/A
This web-based magazine offers general news-like stories related to all aspects of alterative power for home, automobiles and industry.

Manufacturers:

Photovoltaic Panels and Equipment
Photovoltaic panel manufacturers do not supply directly to end consumers, as they rely on their large distribution networks throughout the world. For information purposes, here are some of the major suppliers in this field. Contact the manufacturer or visit their website for distributors in your local area.

AstroPower
Website: www.astropower.com
Phone: 302-366-0400

BP Solar
Website: www.bpsolar.com
Phone: 410-981-0240

Evergreen Solar Inc.
Website: www.evergreensolar.com
Phone: 508-357-2221

Kyocera Solar Inc.
Website: www.kyocerasolar.com
Phone: 800-544-6466

Matrix Solar Technologies
Website: www.matrixsolar.com
Phone: 505-833-0100

RWE Schott Solar
Website: www.asepv.com
Phone: 800-977-0777

Schott Applied Power Corporation
Website: www.schottappliedpower.com
Phone: 888-457-6527

Sharp USA
Website: www.sharpusa.com
Phone: 800-BE-SHARP

Siemens Solar Inc.
Website: www.siemenssolar.com
Phone: 877-360-1789

Solardyne Corporation
Website: www.solardyne.com
Phone: 503-244-5815
Manufacturer of small solar modules for charging laptop computers, cell phones, etc.

Solarex
Website: www.solarex.com
Phone: 301-698-4200
Solarex was recently acquired by BP Solar

PV Module Mounts
Array Technologies Inc.
Website: www.wattsun.com
Phone: 505-881-7567
Manufacturer of the Wattsun active tracking system

Sun-Link Solar Tracker

Website: www.northernlightsenergy.com
Phone: 705-246-2073
Manufacturer of Sun-Link active tracking system.

Two Seas Metalworks

Website: www.2seas.com
Phone: 877-952-9523
Manufacturer of fixed PV racks. Also supply battery racks.

UniRac Inc.

Website: www.unirac.com
Phone: 505-242-6411

Zomeworks Corporation

Website: www.zomeworks.com
Phone: 800-279-6342
Manufacturer of fixed and passive tracking mount systems

Wind Turbines:

Atlantic Orient Corporation

Website: www.aocwind.net
Phone: 802-333-9400

Bergey Windpower Inc.

Website: www.bergey.com
Phone: 405-364-4214

Southwest Windpower

Website: www.windenergy.com
Phone: 520-779-9463

Aeromax Corporation

Website: www.aeromaxwindenergy.com
Phone: 888-407-9463

Bornay Windturbines

Website: www.bornay.com
Phone: +34-965-560-025
Manufacturer of wind turbines from Spain

Lake Michigan Wind and Sun

Website: www.windandsun.com
Phone: 920-743-0456
New and re-built turbines

Jack Rabbit Energy Systems

Website: www.jackrabbitmarine.com
Phone: 203-961-8133

Windstream Power Systems Inc.

Website: www.windstreampower.com
Phone: 802-658-0075

Wind Turbine Industries Corporation

Website: www.windturbine.net
Phone: 952-447-6064

True North Power Systems

Website: www.truenorthpower.com
Phone: 519-793-3290

Windturbine.ca

Website: www.windturbine.ca
Phone: 886-778-5069

Micro Hydro Turbine Systems

Energy Systems and Design

Website: www.microhydropower.com
Phone: 506-433-3151

Jack Rabbit Energy Systems

Website: www.jackrabbitmarine.com
Phone: 203-961-8133

Harris Hydroelectric

Website: www.harrishydro.com
Phone: 831-425-7652

Canyon Industries Inc.
Website: www.canyonindustriesinc.com
Phone: 360-592-5552

HydroScreen Co. LLC
Website: www.hydroscreen.com
Phone: 303-333-6071
Manufacturer of intake screen for micro-hydro systems

Battery Manufacturers

Dyno Battery Inc.
Website: www.dynobattery.com
Phone: 206-283-7450

HuP Solar-One Battery
Website: www.hupsolarone.com
Phone: 208-267-6409

IBE Battery
Website: www.ibe-inc.com
Phone: 818-767-7067

Rolls Battery Engineering (USA)
Surrette Battery Company (Canada)
Website: www.surrette.com
Phone: 800-681-9914

Trojan Battery Company
Website: www.trojanbattery.com
Phone: 800-423-6569

U.S. Battery Manufacturing Company
Website: www.usbattery.com
Phone: 800-695-0945

Hydrogen Recombining Caps

Hydrocap Catalyst Battery Caps
Website: N/A
Phone: 305-696-2504

D.C. Voltage Regulators

Morningstar Corporation
Website: www.morningstarcorp.com
Phone: 215-321-4457

RV Power Products
Website: www.rvpowerproducts.com
Phone: 800-493-7877

Steca Gmbh
Website: www.stecasolar.com
Phone: N/A

Xantrex Technology Inc.
Website: www.xantrex.com
Phone: 360-435-2220

Inverters

ExelTech Inc.
Website: www.exeltech.com
Phone: 800-886-4683

Out Back Power Systems
Website: www.outbackpower.com
Phone: 360-435-6030

SMA America Inc.
Website: www.sma-america.com
Phone: 530-273-4895

Xantrex Technology Inc.
Website: www.xantrex.com
Phone: 360-435-2220

Backup Power Gensets

Epower
Website:
www.epowerchargerboosters.com
Phone: 423-253-6984

Generac Power Systems Inc.
Website: www.generac.com
Phone: N/A (sold through Home Depot)

Hardy Diesel & Equipment Inc.
Website: www.hardydiesel.com
Phone: 800-341-7027

Kohler Power Systems
Website: www.kohlerpowersystems.com
Phone: 800-544-2444

Energy Meters

Bogart Engineering
Website:
www.borartengineering.com
Phone: 831-338-0616

Brand Electronics
Website: www.brandelectronics.com
Phone: 207-549-3401

Xantrex Technology Inc.
Website: www.xantrex.com
Phone: 360-435-2220

Miscellaneous

Bussmann
Website: www.bussmann.com
Phone: 314-527-3877
Fuses and electrical safety components

Delta Lightning Arrestors Inc.
Website: www.deltala.com
Phone: 915-267-1000

Digi-Key (Canada and USA)
Website: www.digikey.com
Phone: 800-DIGI-KEY
Supplier of many electrical wiring components

Electro Sonic Inc. (Canada)
Website: www.e-sonic.com
Phone: 800-56-SONIC
Supplier of many electrical wiring components

Real Goods
Website: www.realgoods.com
Phone: 800-919-2400
Suppliers specializing in renewable energy systems

Siemens Energy and Automation Inc.
Website: www.siemens.com
Phone: 404-751-2000
Fused disconnect switches and circuit breakers

Xantrex Technology Inc.
Website: www.xantrex.com
Phone: 360-435-2220
D.C. circuit breakers, battery cables, power centers, metering shunts, fuses

Appendix 4
Magnetic Declination Map for North America

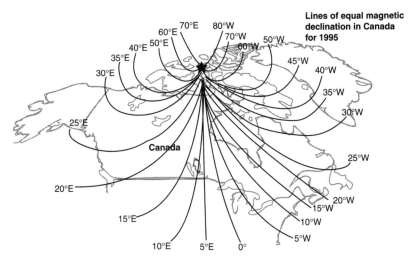

Lines of equal magnetic declination in Canada for 1995

Map of US showing magnetic declatination for Azimuth compass calculations

Subtract from Azimuth Calculation ◄──► Add from Azimuth Calculation
for Compass Heading   for Compass Heading

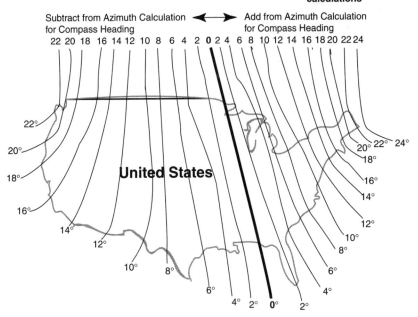

Map indicates the correction heading required to locate true north (and solar south). If you live along the line running from Manitoba through to southern Texas, your heading will be due north when the compass is pointing approximately 8 degrees east.

Appendix 5
Winter Average Sun Hours per Day Map for North America

"North American Sun Hours per Day (Worst Month)"

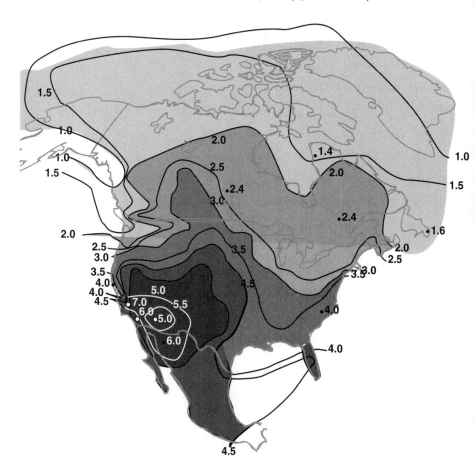

Map indicates the winter average of sun hours per day. Use this map when calculating average energy output of a PV system used all year round.

Appendix 6
Yearly Average Sun Hours per Day Map for North America

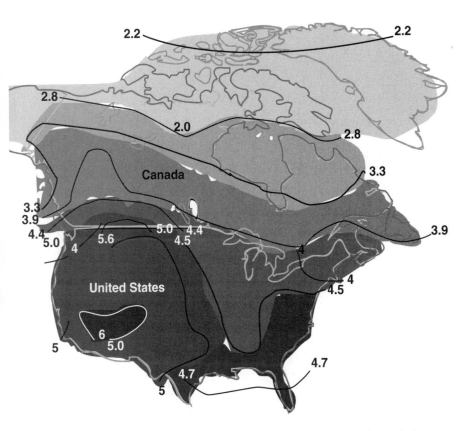

Map indicates the yearly average of sun hours per day. Use this map when calculating average energy output of a PV system used seasonally during the spring/summer months.

Appendix 7
Electrical Energy Consumption Worksheet

Appliance Type	Appliance Wattage (Volts x Amps)	X	Hours of Daily Use	=	Average Watt-hours Per Day

Total Watt-hours Per Day of all Appliances = [   ]

Start by checking your major electrical loads in the house. Anything that plugs into a regular wall socket will have a label that tells you the voltage (usually 120 or 240 for larger appliances) and current or wattage for that item. Although these labels can over-state energy usage, they are a good guide for calculating energy consumption. Write down the wattage data from the labels into the *Energy Consumption Worksheet*. (If the appliance label does not show wattage, multiply amps x volts to calculate watts).

The next step is to see if you can estimate how many hours you use the device per day. If you only use a device occasionally, try to estimate how long it is used per week and divide this time by 7. A quick fly through the calculator and voila', your "daily average energy usage per day" is computed.

Appendix 8
Average Annual Wind Speed Map for North America

Wind Class / Speed Chart

Class 1:   3.8 m / s (8.5 mph)
Class 2:   4.8 m / s (10.8 mph)
Class 3:   5.4 m / s (12.1 mph)
Class 4:   5.8 m / s (13.0 mph)
Class 5:   6.2 m / s (13.9 mph)
Class 6:   6.7 m / s (15.0 mph)
Class 7:   7.5 m / s (16.8 mph)

Map indicates the average annual wind speed in meters per second averaged over a ten-year period. Elevation of anemometer at 33 feet (10 m). Use this map with care as local wind speed levels will vary. Refer to resource guide for further details on wind mapping for your specific area.

Appendix 9
Voltage, Current and Distance (in feet) Charts
for Wiring (AWG gauge)
(1% Voltage drop shown – Increase distance proportionally if greater losses allowed)

12 Volt Circuit

Amps	WIRE GAUGE									
	10	8	6	4	2	1	1/0	2/0	3/0	4/0
1	48	74	118	187	299	375	472	594	753	948
5	10	15	24	37	60	75	94	119	151	190
10	5	7	12	19	30	38	47	59	75	94
20	2	4	6	9	15	19	24	30	38	48
40	1	2	3	5	7	9	12	15	19	24

24 Volt Circuit

Amps	WIRE GAUGE									
	10	8	6	4	2	1	1/0	2/0	3/0	4/0
10	10	15	24	37	60	75	94	119	151	190
20	5	7	12	19	30	38	47	59	75	94
30	3	5	8	12	20	25	31	40	50	63
40	2	4	6	9	15	19	24	30	38	48
50	2	2	4	6	10	13	16	20	25	31
100	1	1	2	4	6	8	9	12	15	19
125	1	1	2	3	5	6	8	10	13	16
150	1	1	2	2	4	5	6	8	10	13

48 Volt Circuit

Amps	WIRE GAUGE									
	10	8	6	4	2	1	1/0	2/0	3/0	4/0
10	20	30	48	74	120	150	188	238	302	380
20	10	15	24	38	60	76	94	118	150	188
30	6	10	16	24	40	50	62	80	100	126
40	4	8	12	18	30	38	48	60	76	96
50	4	4	8	12	20	26	32	40	50	62
100	2	2	4	8	12	16	18	24	30	48
125	2	2	4	6	10	12	16	20	26	32
150	2	2	4	4	8	10	12	16	20	26

Appendix 10
Wire Sizes Versus Current Carrying Capacity
(Size shown is for copper conductor only)

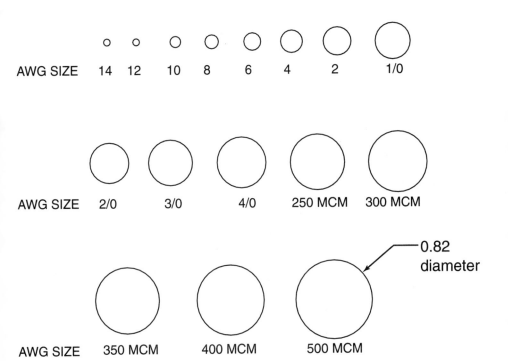

| AWG SIZE | 14 | 12 | 10 | 8 | 6 | 4 | 2 | 1/0 |

| AWG SIZE | 2/0 | 3/0 | 4/0 | 250 MCM | 300 MCM |

0.82 diameter

| AWG SIZE | 350 MCM | 400 MCM | 500 MCM |

AWG	Maximum Current
14	15
12	20
10	30
8	55
6	75
4	95
2	130
0	170
2/0	195
3/0	225
4/0	260

Index